VIBRATORY PILE DRIVING AND DEEP SOIL COMPACTION-TRANSVIB2002

PROCEEDINGS OF THE INTERNATIONAL CONFERENCE ON VIBRATORY PILE DRIVING AND DEEP SOIL COMPACTION / LOUVAIN-LA NEUVE / BELGIUM / 9-10 SEPTEMBER 2002

# Vibratory Pile Driving and Deep Soil Compaction -TRANSVIB2002

*Edited by*

A. Holeyman
*Université catholique de Louvain (UCL), Belgium*

J-F. Vanden Berghe
*THALES GeoSolutions, Belgium*

N. Charue
*Université catholique de Louvain (UCL), Belgium*

A.A. BALKEMA PUBLISHERS  LISSE / ABINGDON/EXTON (PA) / TOKYO

Published by: A.A. Balkema, a member of Swets & Zeitlinger Publishers
www.balkema.nl and www.szp.swets.nl


ISBN 90 5809 521 5

Printed in the Netherlands

*Vibratory Pile Driving and Deep Soil Compaction - TRANSVIB2002,*
*Holeyman, VandenBerghe & Charue (eds.), © 2002 Swets & Zeitlinger, Lisse, ISBN 90 5809 521 5*

# Table of Contents

Preface     IX

Acknowledgements     XI

Organization     XIII

## *Keynote Lectures*

Soil behavior under vibratory driving     3
*A. Holeyman*

Modeling of vibratory pile driving     21
*F. Rausche*

Effects of vibratory compaction     33
*K. R. Massarsch*

## *Modeling of vibratory driving*

A mechanical model for the investigation of the vibrodrivability     45
of piles in cohesionless soils
*R. O. Cudmani, G. Huber & G. Gudehus*

Vibratory pile driving analysis - a simplified model     53
*J.-G. Sieffert*

Application of a hypoplastic constitutive law into a vibratory pile driving model     61
*J-F. Vanden Berghe & A. Holeyman*

Simple model for prediction of vibratory driving and experimental     69
data analysis of vibro-driven probes an sheetpiles
*D. Vié*

# Case histories and instrumented vibratory driving

Sheet Pile Driving & Testing M5 East Project Sydney, Australia     83
*S. Baycan*

Drivability prediction of vibrated steel piles     89
*N. Huybrechts, C. Legrand & A. Holeyman*

Vibrodrivability and induced ground vibrations of vibratory installed sheet piles     99
*K. Viking*

# Deep soil vibrocompaction

On-line compaction control in deep vibratory compaction     115
*W. Fellin, G. Hochenwarter & A. Geiß*

Foundation of wind power station on mining dump applying     123
deep soil compaction
*Y. El-Mossallamy, J. Löschner, W. Kissel & B. Schlesinger*

On- and offshore vibro-compaction for an oil pipeline in Singapore     129
*W.C.S. Wehr & V. R. Raju*

# Environmental aspects of vibratory operations

A case study on safe sheet pile driving with vibration monitoring     135
*W. Haegeman*

Settlement due to sheetpile extraction, results of experimental research     141
*P. Meijers & A.F.van Tol*

Construction process induced vibrations on underground structures     147
*I.Thusyanthan & S.P.G. Madabhushi*

# Soil investigation and bearing capacity

Development of a Vibro-Penetration Test (VPT) for insitu investigation     157
of cohesionless soils
*R. O. Cudmani & G.Huber*

Two comparative field studies of the bearing capacity of     167
vibratory and impact driven sheet piles
*S. Borel, M. Bustamante & L. Gianeselli*

Pile load tests of large diameter battered steel pipe piles     175
constructed on the offshore area
*J. H. Lee, D. D. Seo, S. Y. Kwon, J. H. Won & Y. I. Baek*

## Vibratory prediction event: *Event and predictions comparison*

Full-scale behaviour of vibratory driven piles in Montoir    179
*S. Borel,L. Gianeselli, D. Durot, P. Vaillant, L. Barbot, B. Marsset & P. Lijour*

International prediction event of vibratory pile driving    193
*J-F. Vanden Berghe*

## Vibratory prediction event: *Individual contributions*

Vibratibility Predictions using the β-Method    213
*G.Jonker & N. van der Zouw*

Class A prediction of the vibrodrivability of a steel pile in Montoir    219
*R. O. Cudmani*

Vibrodrivability of piles at Montoir prediction using a simplified model    225
*J.-G. Sieffert*

Author index    233

# Preface

The application of vibratory driving techniques to solve civil engineering challenges appears to date back to the early 30's. The observation by early researchers that soil resistance could be reduced thanks to vibrations led to the industrial use of vibrators to drive piles. Extensive research on the effects of vibration on soils was conducted in the 40's and 50's, while the vibratory driving technique was gaining acceptance as an economical and effective means of installing piles and sheet piles in appropriate soil conditions. Although technological developments have been brought to enhance the initial concept and extend commercial application of vibrators, little is mastered by the engineer when it comes to addressing soil related issues. That limitation in the engineering knowledge is viewed by many as an impediment for the vibratory driving technique to enjoy its full potential.

A need was therefore felt to bridge the gaps between the various bodies of knowledge that existed throughout the world in that field. The format of an international conference was identified as an ideal platform to bring closer researchers, equipment manufacturers, and operators. Thanks to the enthusiasm and insistence of younger researchers of the Geotechnical Research Group ("GeoMEM") at the Catholic University of Louvain ("UCL"), Louvain-la-Neuve was suggested to the international engineering community as the meeting place for the very first gathering of that type. An organizing committee was assembled in 2000, which helped define the themes of the conference as "Vibratory Pile Driving and Deep Soil Compaction".

A number of favourable factors legitimated the venue in Belgium, Europe. The Belgian Building Research Institute ("BBRI") had indeed stimulated and taken part into two major research programs during the early 90's dealing with vibratory driving (HYPERVIB1 and HYPERVIB2), financed by the European Union. It is through those programs that I started getting interested in that most fascinating problem. Research in the field has since then been pursued at Louvain-la-Neuve. The near-by site of Montoir, in France, selected by the IREX National project to perform vibratory penetration tests, also offered a unique opportunity to organize a Class-A prediction event.

How much more challenging can a geotechnical problem be, yet with simple geometry? It is because the interaction of soil with a strongly vibrating structure involves dynamic, cyclic, non-linear, as well as material transformation aspects that have been rarely assembled into a situation that is routinely encountered on many civil engineering projects. Clearly, fundamental understanding lags operational application in vibratory driving and compaction.

Besides keynote lectures on soil behavior, modeling of vibratory pile driving, and vibratory compaction, the several contributions included herein combine to provide a wide treatment of the subject of vibratory driving and deep soil compaction, including:
- Modeling of Vibratory driving
- Case histories and instrumented vibratory driving
- Deep soil vibro-compaction
- Environmental aspects of vibratory operations
- Soil investigation and bearing capacity

A special section is devoted to the vibratory prediction event made possible through cooperation of the Conference Organizing Committee with the French National Project "Vibro-fonçage" initiated by IREX. The observations and measurements during the tests as well as the comparison of predictions are documented. Class-A predictions and "post-dictions" are also included on an individual basis to stimulate discussion.

An accomplishment of that magnitude would not have been possible without the collective effort of the team assembled for that occasion. The comprehensive list of individuals and organizations to be thanked for their contributions under many forms is too vast to be included in this preface and is therefore provided under a specific Acknowledgement section. I would like however to single out the financial support from the National Foundation for Scientific Research ("FNRS.") which enabled the GeoMEM group to pursue research on the fundamental understanding of soil behavior under high-strain cyclic loading and which provided funds to support the organization of this highly specialized and unique international conference. It is also my pleasure to specially thank Jean-François Vanden Berghe and Nicolas Charue for their enthusiastic support in organizing the conference.

As Chairman of the Organizing Committee and Editor of this volume, it is my hope that the material provided herein will improve both fundamental and practical understanding necessary to master as well as expand the international confidence towards the application of vibratory techniques to pile driving and deep soil compaction.

Alain E. Holeyman
Chairman of the Organizing Committee
Louvain-la-Neuve, Belgium

# Acknowledgements

The Editors wish to express their warmest gratitude to the following parties:

*for the financial support of the conference :*
- Le Fonds National de la Recherche Scientifique, Belgium.

*for the organization of the prediction event:*
- The IREX project (French national research project on drivability and bearing capacity of vibratory driven piles) for the organization of the test,
- All the predictors, for their effort and intellectual courage.

*for the organization of the conference :*
- The Conference Organizing Committee,
- The International Scientific Committee,
- The Contributions Reviewers,
- The Secretariat of Civil and Environmental Engineering Department of the Université catholique de Louvain,
- The Public Relation Department of the Université catholique de Louvain.

*for the sponsoring of the conference :*
- ISPC Company for its support for the banquet and the lunches of the conference,
- PDI Company for its support for the coffee breaks of the conference.

*for the technical exhibit, the exhibitors :*
- Dieseko's Piling & Vibro Equipment (PVE),
- International Construction Equipment (ICE),
- Profound BV,
- PTC,
- IHC.

*Vibratory Pile Driving and Deep Soil Compaction - TRANSVIB2002,*
*Holeyman, VandenBerghe & Charue (eds.), © 2002 Swets & Zeitlinger, Lisse, ISBN 90 5809 521 5*

# Organization

## CONFERENCE ORGANIZING COMMITTEE

*Chairman:*

Prof. Dr. Ir. A. Holeyman, Université catholique de Louvain (UCL), Belgium

*Secretary:*

Dr. Ir. J-F. Vanden Berghe, THALES GeoSolutions, Belgium

*Treasurer:*

Ir. N. Charue, Université catholique de Louvain (UCL), Belgium

*Members:*

Dr. Ir. S. Borel, Laboratoire Central des Ponts et Chaussées (LCPC), France

Prof. Dr. Ir. G. Degrande, Katholieke Universiteit Leuven (KUL), Belgium

Prof. Dr. Ir. W.Haegeman, Universiteit Gent (UG), Belgium

Ir. N. Huybrechts, Belgian Building Research Institute (BBRI), Belgium

Ir. C. Legrand, Belgian Building Research Institute (BBRI), Belgium

Dr. Ir. A. Schmitt, ISPC, Luxemburg

Ir. G. Simon, Walloon Ministry of Transportation (MET), Belgium

Prof. Dr. Ir. J.-C. Verbrugge, Université Libre de Bruxelles (ULB), Belgium

Dr. Ir. K. Viking, KTH, Sweden

# CONFERENCE SCIENTIFIC COMMITTEE

Ir. H. Gonin, France

Prof. Dr. Ir. A. Holeyman, Belgium

Prof. Dr. Ir. K. Ishihara, Japan

Dr. Ir. F. Rausche, USA

Prof. Dr. Ir. M. Randolph, Australia

Prof. Dr. Ir. W. Van Impe, Belgium

# PAPER REVIEWING COMMITTEE

The editors are grateful to the following persons who helped to review the manuscripts and thus greatly assisted in improving the overall technical standard and presentation of the papers in these proceedings :

Prof. Dr. Ir. G. Degrande, Katholieke Universiteit Leuven (KUL), Belgium

Prof. Dr. Ir. W.Haegeman, Universiteit Gent (UG), Belgium

Dr. Ir. K. Viking, KTH, Sweden

Dr. Ir. S. Borel, Laboratoire Central des Ponts et Chaussées (LCPC), France

*Keynote lectures*

# Soil Behavior under Vibratory Driving

Alain E. Holeyman

*Université Catholique de Louvain, Louvain-la-Neuve, Belgium*

ABSTRACT : Vibratory driving of piles and soil densification raise several engineering issues including the long-term bearing capacity of the installed pile, its vibratory penetration resistance, the performance of vibrators, the degradation and liquefaction of the soil around the vibrated profile, and vibratory nuisance to the environment. An argument is made that those issues will be adequately tackled once combined into a comprehensive framework of analysis where proper understanding of soil behavior is the key. The present paper reviews our current engineering ability to assess vibro-drivability, i.e. predicting the vibratory penetration log of a given pile into a given soil profile using a given vibrator. Testing undertaken to provide insight into the pile-soil-vibrator interaction and its modelling is then emphasized. Several available methods to establish the vibratory performance of a pile from its vibratory capacity are discussed. A rational procedure to model the dynamic nonlinear soil structure interaction during pile vibratory driving is discussed in more detail. Degradation of the skin friction upon cyclic shear stress is evaluated by applying elements of earthquake engineering practice used to assess liquefaction potential. The present ability to assess the vibratory capacity of a pile from the monitoring of its vibratory performance is critically reviewed. Finally, suggestions for further research, design and practice are provided.

## 1 INTRODUCTION

### 1.1 *Scope*

The main purpose of this key-note paper is to present the author's present view on soil engineering issues relating to the drivability of piles and sheet-piles using vibrators and the inverse problem, i.e. deriving the pile resistance from its vibratory performance. It is based on a survey of the relevant literature and original research in the area.

### 1.2 *Historical Development*

The vibratory driving technique appears to date back to the early 30's when it was simultaneously developed in the former USSR and in Germany (Rodger and Littlejohn, 1980). The observation by the Russian soil dynamics researcher Pavyluk that soil resistance could be reduced thanks to vibrations led to the industrial use of vibrators to drive piles, according to Barkan, 1960. Extensive research on the effects of vibration on soils was conducted in the 40's and 50's by Barkan, while the vibratory driving technique was gaining acceptance as an economical and effective means of installing piles and sheet piles in appropriate soil conditions. Major vibrators manufacturers are now located in Germany, France, The Netherlands,

USA, the Former Soviet Union and Japan. Although technological developments have been brought to enhance the initial concept and extend commercial application of vibrators, little is mastered by the engineer when it comes to addressing soil related issues. That limitation in the engineering knowledge is viewed by many as an impediment for the vibratory driving technique to enjoy its full potential.

### 1.3 *Phenomena at Play*

Three major actors play a role in the mechanics of the vibratory driving process, as illustrated in Fig. 1 : (1) the pile to be driven, (2) the selected vibrator, and (3) the imposed soil conditions. The pile can be fully described by its material and geometry. The vibrator mechanical behavior can be assessed based on its specifications and operational range, as discussed in Section 2. Soil conditions are usually characterized by means of standard investigation tools such as CPT soundings, borings and laboratory tests.

Those investigation tools are geared towards answering general design questions (mostly static) but are not well suited to characterize soil behavior under pile installation conditions, specially if the piles are vibratory driven.

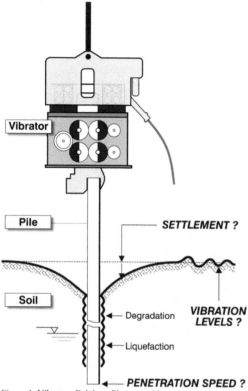

Figure 1. Vibratory Driving : Players and Issues

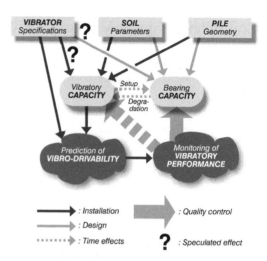

Figure 2. Installation and design process of vibratory driven piles

Because it is has been established for more than half a century that soil resistance during vibratory driving (likewise during impact driving) is lower than the long-term bearing capacity, these two resistances should be distinguished. As shown in Fig. 2, one can estimate the vibratory capacity from the long-term bearing capacity by taking soil degradation effects into account. Conversely, one can estimate the long-term bearing capacity from the vibratory capacity if soil set-up can be accounted for.

A fundamental understanding of soil behavior under vibratory loading is required to establish the relationship between the pile vibratory resistance and its long-term bearing capacity. Soil resistance degrades upon cyclic shearing mainly because of fatigue of the soil skeleton in cohesive soils (Vucetic, 1992), and of effective stress reduction in granular soils (Casagrande, 1938). The effective stress can be ultimately reduced to nearly zero, at which point the soil behaves in a fluid-like manner. These phenomena will be reviewed in more detail in Section 3.

## 1.4 Engineering issues

Engineering issues related to vibratory driving cover many facets, as illustrated in Figs. 1 and 2. They prompt the following questions :

- What is the long-term bearing capacity of the installed pile? We know it depends on the pile geometry, on the soil parameters, but also on the vibratory process.
- How are the soil's long-term strength parameters influenced by the vibratory process? By how much will the soil compact, and what is the magnitude of the potentially induced settlement?
- How are vibrations transmitted to the surrounding soil, and how much potential damage can they cause to neighboring structures?
- Will a given vibrator be able to drive the pile to the required design depth? If so, at what speed? Are there soil types that strongly limit vibratory penetration depth?
- Are there ways to assess the vibratory capacity of a pile from the monitoring of its vibratory performance?
- Is there a vibratory testing technique and interpretation leading to estimating the long-term pile bearing capacity?

One can actually state that all issues will be properly tackled when combined into a comprehensive framework of analysis where proper understanding of soil behavior is the key. As clarified in Fig. 2, the present paper will focus on our current engineering ability to assess vibro-drivability, i.e. to predict the vibratory penetration log of a given pile into a given soil profile using a given vibrator. Testing undertaken to provide insight into the pile-soil-vibrator interaction and its modelling will be reviewed in Sections 4 and 5. We will then focus on some available methods to establish the vibratory performance of a pile from its vibratory capacity (Section 6), and look into the potential to establish the reverse relationship (Section 7). Finally, Section 8 provides suggestions for further research, design and practice.

Because of space limitations, the paper does not focus on other important engineering issues such as: bearing capacity of vibro-driven piles derived from soil characterization, vibrations transferred to the environment, and equipment specifications.

## 2 PILES AND VIBRATORY EQUIPMENT

### 2.1 Vibrated Piles

Pile types or profiles mostly used in combination with the vibratory driving technique include:
- sheet piles installed for temporary shoring, cofferdam and permanent retaining and containing walls,
- H-piles vibro-driven as deep foundations or vibrated to help install underground hydraulic barriers,
- Tubes to install cast-in-steel-shell (CISS) piles
- Precast prestressed concrete piles
- Steel profiles to vibro-compact granular soils at depth.

Port, harbor, near-shore and offshore projects very often take advantage of the vibratory penetration techniques, as the environment lends its self to substantial tolerance of vibratory disturbances.

Steel and concrete profiles are generally cylindrical or prismatic, and can be characterized by the following geometrical and mechanical properties :
A [m²]: profile section
L [m]: profile length
χ [m]: profile perimeter
E [Mpa]: Material Young's Modulus
ρ [kg/m³]: Material volumic mass

The profile section can be more fully characterized by its shape, and inside and outside perimeters if closed. This allows one to calculate the areas of the profile in longitudinal and transversal contact with the soil, once an embedment depth z [m] is assumed. The mass of the profile $M_p$ equals $\rho AL$ [kg] while the longitudinal wave speed in the profile is given by $c=\sqrt{E/\rho}$ $[m/s]$.

Although they may at time play an important role, transversal and flexural properties of the profile will generally be ignored in the analysis that confines itself to the longitudinal behavior of the profile.

### 2.2 Mechanical action of a vibrator

The mechanical action of a vibrator onto a profile consists of two part: a vibratory action and a stationary action.

The vibratory action imparted to the pile is produced by counter-rotating eccentric masses actuated within an "exciter block", as shown in Fig. 3a. The centrifugal forces acting as a result of inertial effects on an even number of symetrically moving masses

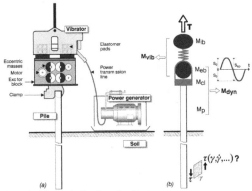

Figure 3. Mechanical action of a vibrator

combine into a sinusoidal vertical force :

$$F_v(t) = me \cdot \omega^2 \sin(\omega t) = F_c \sin(\omega t) \qquad (1)$$

where
$F_c$ = maximum centrifugal force of the vibrator [N]
me = eccentric moment of the vibrator [kg.m]
ω = angular frequency of the vibrator [rad/s]

Alternative quantifications of the angular frequency are the rotation speed R [rpm] and the frequency ν [Hz], with :

$$R = 60\,v = 60 \cdot (\omega/2\pi) \qquad (2)$$

The vibratory action can be therefore assessed once both the eccentric moment and the operating frequency are known. That action will be balanced by reactive inertial effects of masses undergoing the imparted vibratory movement and by soil reactions opposing the profile movement. Provided the center of gravity of the rotating masses belongs at all times to the profile neutral axis, the exciter block is assumed to exert a purely longitudinal force onto the profile.

The exciter block is connected to the profile via a clamping device and is suspended to a carrier. The suspension device includes a vibration isolator mechanism consisting of a quasi-stationary heavy mass directly suspended to the suspension hook and an intervening spring, generally consisting of elastometer pads.

The vibrator can be viewed as a two degrees of freedom system moving in the longitudinal direction (see Fig. 3b) : an exciter block of mass $M_{eb}$ and an isolator block of mass $M_{ib}$, sometimes called bias mass. Therefore $M_{vib} = M_{eb} + M_{ib}$. Those two masses are interconnected via an isolation spring with constant $k_i$. In addition to the effort generated by that spring, the mass $M_{eb}$ is subjected to gravity (g) and the sinusoidal force described by eq. (1) whereas the mass $M_{ib}$ is subjected to gravity and the suspension force T. The net quasi-stationary action on the pile resulting from the carrier operation and vibrator is the weight of the vibrator mass and its clamp $M_{cl}$ deducted by the suspension force:

5

$$Fs\ [N]\ =\ (M_{vib}\ +\ M_{cl})\ .\ g - T \qquad (3)$$

## 2.3 Vibrator Movement

The movement of the vibrated body will depend on its so-called dynamic mass and the soil resistance. Specifications of vibrators often list a "maximum amplitude" $S_{sp}$. That number [generally expressed in mm] corresponds to the *total* (i.e. double) amplitude of movement for a free hanging vibrator, thus assuming a dynamic mass consisting of the exciter block $M_{eb}$ and the clamping device $M_{cl}$:

$$S_{sp}\ =\ 2\ s_0\ =\ 2\ me/\ (M_{eb}\ +\ M_{cl}) \qquad (4a)$$

It should be noted that the double amplitude does not depend on the operating frequency, as the center of mass of the free mechanical system remains stationary, irrespective of the frequency. The amplitude of the free hanging pile to be vibrated will always be smaller than the specified amplitude, as can be derived from eq. (4a), where the dynamic mass is increased by that of the pile ($\rho AL$).

$$2s = S_{sp} \cdot \frac{M_{eb} + M_{cl}}{M_{eb} + M_{cl} + M_p} \qquad (4b)$$

with s = actual (single) amplitude of the dynamic mass.

A power for the vibrator is often listed in the specifications. It generally corresponds to the nominal power of the motor actuating the eccentric masses. It does not correspond to standardized operational conditions of the vibrator in action. Power consumption is indeed dependent upon testing conditions. Barkan suggests that under pile vibratory conditions, the power follows a squared velocity law:

$$W\ [kW]\ =\ M_{dyn}\ .\ n\ .\ (s\ \omega)^2 = \beta_t \cdot W_t \qquad (4c)$$

Experimental verification of that law shows the n value to depend on soil type and pile type; a range of 15 to 50 Hz is observed.
O'Neill and Vipulanandan (1989) provide an expression of the theoretical power required to maintain the vibrating regime of a dynamic mass in the absence of soil reaction but accounting for the presence of the isolating spring and the bias mass $M_{ib}$. That formula is however of limited practical use as it provides very low estimates of the power.

It is the author's opinion that power limitation of the equipment is neither sufficiently characterized, nor (therefore?) properly accounted for in vibratory driving analyses conducted to date.

## 2.4 Types of vibrators

Two main types of vibrators are commercially available: hydraulic and electrical. In both cases, the motor is housed in the vibrator and powered through a transmission line connected to a separate or carrier-mounted diesel-hydraulic or diesel-electric power pack (see Fig. 3a). Hydraulic vibrators are lighter than their electrical counterparts, because of the smaller size of the motor. The adjustment of the operating frequency is more readily available on the hydraulic vibrators, which also explain why they are more commonly used.

Five types of vibrators can be distinguished based on operating frequency and eccentric moments, as summarized in Table 1.

It can be noted that initial improvements of the vibratory driving technique targeted the speed of driving, whereas more recent improvements are attempting to mitigate environmental impacts associated with the technique. Noteworthy amongst recent developments is the "variable vibrator", which can adjust on-the-fly its effective eccentricity by shifting the phase angle between a multiple of 4 masses. The claimed advantage of such an adjustment is to avoid "soil resonance", a term coined after the observation that vibration levels pass through a peak upon vibration start-up and shut-down. This phenomenon will however be shown later not to be necessarily related to a particular frequency.

Vibrator choice amongst practicionners is generally based on experience and field verification. Rodger and Littlejohn (1980) have summarized that body of experience into a table recommending frequency and amplitude parameters for different piles and soil types. Those recommendations are reproduced herein as Table 2.

Table 1. Vibrator types

| Type | Frequency range [rpm] | Eccentric moment [kg.m] | Maximum centrifugal force [kN] | Free hanging double amplitude [mm] |
|---|---|---|---|---|
| "Standard frequency" | 1300-1800 | up to 230 | up to 4,600 | up to 30 |
| High frequency | 2000-2500 | 6 to 45 | 400 to 2,700 | 13 to 22 |
| Variable eccentricity | 2300 | 10 to 54 | 600 to 3300 | 14 to 17 |
| Excavator accessory | 1800 to 3000 | 1 to 13 | 70 to 500 | 6 to 20 |
| Resonant driver | 6000 | 50 | 20,000 (in theory) | Self destructing |

Table 2. Vibrators classification (after Rodger and Littlejohn, 1980)

| Cohesive soils | Dense cohesionless soils | | Loose cohesionless soils | |
|---|---|---|---|---|
| All cases | Low point resistance | High point resistance | Heavy piles | Light piles |
| High acceleration Low displacement amplitude | High acceleration | Low frequency. Large displacement amplitude | | High acceleration |
| Predominant side resistance | Predominant side resistance. | Predominant end resistance. | | Predominant side resistance. |
| Requires high acceleration for either shearing or thixotropic transformation | Requires high acceleration for fluidization | Requires high displacement amplitude and low frequency for maximum impact to permit elasto-plastic penetration | | Requires high acceleration for fluidization |
| Recommended parameters | | | | |
| ν > 40 Hz a: 6-20 g s : 1-10 mm | ν : 10-40 Hz a : 5-15 g s : 1-10 mm | ν : 4-16 Hz a: 3-14 g s : 9-20 mm | | ν : 10-40 Hz a : 5-15 g s : 1-10 mm |

# 3 SOIL BEHAVIOR UNDER VIBRATORY LOADING

## 3.1 Fundamentals

As the profile undergoes a vibratory vertical motion of amplitude s, it communicates to the lateral neighboring soil shear stresses and shear strains, as sketched in Fig. 3b. It is also forcing normal and po-tentially convective movement of soil below the pile toe.As those mechanisms govern soil resistance along the shaft and at the toe, the understanding of the shear stress/shear strain relationship, i.e. $\tau$ ($\gamma$), within the soil becomes of paramount importance.

That aspect of soil behavior has been more extensively studied within the field of earthquake engineering, leading to the characterization of so-called constitutive relationships, generally on the basis of laboratory testing of soil samples (mainly triaxial testing and simple shear testing). The constitutive relationships that represents the complex large-strain, dynamic and cyclic shear stress-strain strength, behavior of the medium surrounding the vibrating profile require the characterization of the following elements :

- Static stress-strain law expressing nonlinear behavior under monotonic loading and hysteresis upon strain reversal,
- Shear modulus at small strains and ultimate shear strength,
- Softening and increase of hysteretic damping with increasing strain,
- Effect of strain rate on initial shear modulus and ultimate strength,
- Degradation of properties resulting from the application of numerous cycles, and last but not least,
- Generation of excess pore pressure leading substantial loss of resistance and possibly to liquefaction.

The following paragraphs address key components of the constitutive relationships and provide insight on the intrinsic soil behavior in the vicinity of the vibrating profile.

## 3.2 Static and Cyclic Stress-strain Behavior

A typical soil response to uniform cyclic strains with amplitude $\gamma_c$ is represented in Fig. 4, which highlights the following fundamental parameters:
$G_{max}$ : initial (or tangent) shear modulus
$\tau_c$ : shear stress mobilized at $\gamma_c$
$G_s$: secant (or equivalent) shear modulus
$\lambda$ : hysteretic (or intrinsic) damping ratio;

$$\lambda = \Delta W / 2\pi\gamma_c \tau_c \qquad (5)$$

with $\Delta W$ = Energy lost during a given cycle.

Both $G_s$ and $\lambda$ are strain-dependent parameters that need to be described by specific laws within a given cycle. $\tau_{max}$ is the ultimate shear strength, revealed at large strains. $\tau_{max}$ and $G_{max}$ are shown to decrease with the number of cycles (cyclic degradation).

## 3.3 Initial Shear modulus and ultimate shear strength ($G_{max}$ and $\tau_{max}$)

Numerous studies have dealt with the initial shear modulus to be used in earthquake engineering (e.g. Drnevich et al., 1967). Most of them are supported by parameters determined in the laboratory which are generally not available when a vibratory penetration issue arises. However, correlatins with CPT test results have been more recently developed (Seed and De Alba, 1986, Robertson and Wride, 1998)

## 3.4 Secant Shear Modulus and Hysteretic Damping ($G_s$ and $\lambda$)

As can be observed in Fig. 4, $G_s$ decreases with the shear strain during the initial monotonic loading. The curve that represents the initial monotonic loading is referred to as the initial "backbone" curve, because it also serves as the basis to generate the

7

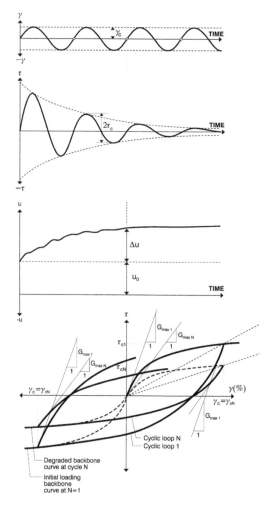

Figure 4. Soil Behavior under Constant Cyclic Shear Strain Amplitude Loading (From Vusetic, 1993; 1994)

family of curves corresponding to unloading and reloading. Kondner's mathematical formulation (1963) is frequently employed to describe the initial backbone curve in earthquake engineering. That hyperbolic law is best represented in terms of reduced variables, $\eta$, the mobilization ratio and $\delta$, the relative shear :

$$\eta = \tau / \tau_{max} = \delta /(\delta + 1)$$

with $\quad \delta = \gamma/\gamma_r = \gamma \cdot G_{max}/\tau_{max} \quad$ (6)

and $\quad \gamma_r = \tau_{max}/G_{max}$

$\gamma_r$ is called the reference strain. Two of the three parameters $G_{max}$, $\gamma_r$, and $\tau_{max}$, are generally derived from laboratory experiments. More extensive labora-

tory surveys by Robertson and Wride (1998) point towards the upward curvature of the stress/strain curve at large cyclic strains.

From the point of maximum straining, the unloading curve is described by the following equation, in accordance with Masing's rules 1 and 2 (Masing, 1926):

$$\tau = \tau_0 (\gamma - \gamma_0)/(1/G_{max} + (\gamma - \gamma_0)/2\tau_{max}) \quad (7)$$

The energy dissipated within a loop depends for a given soil on the amplitude of the cyclic strain. Empirical data collected in laboratory tests indicates that the damping ratio increases with $\gamma_c$ as the soil undergoes higher plastic deformations.

Dobry and Vucetic (1987, Vucetic and Dobry, 1991, and Vucetic,1993 and 1994) have suggested a unifying approach to accommodate the influence of the nature of the material characterized by the plasticity index (PI), as indicated in Fig. 5

### 3.5 Strain Rate Effects

Although it is well known that undrained modulus and shear strength increase with increasing strain rate ($\dot{\gamma} = \partial \gamma / \partial t$), experimental data generated using different apparatuses and loading conditions lead to different conclusions. Viscosity mechanisms may well provide a suitable framework to understand the strain rate effect observed when comparing fast and slow undrained monotonic stress-strain curves, as well as to explain the roundness of the loop tips during a sinusoidal strain-controlled cyclic test. Evidence would point to the fact that sands and non plastic silts have very small viscosity in that their stress-strain loops exhibit sharp rather than rounded tips (Dobry and Vucetic, 1987).

The mathematical functions proposed in the literature to represent the nonlinear viscosity also depend on the type of experimental observations. A power law is often adopted :

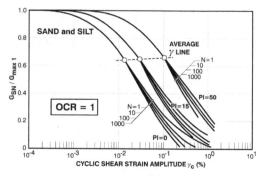

Figure 5. Soil stiffness degradation resulting from cyclic shear (Vucetic, 1993)

8

$$\tau_{kin} = \tau_{sta} \cdot \left(1 + J \cdot \dot{\gamma}^n\right) \qquad (8)$$

with  $\tau_{kin}$ = kinetic ultimate shear strength [kPa]
 $\tau_{sta}$ = "static" ultimate shear strength [kPa]
 $\dot{\gamma}$ = shear strain rate [s⁻¹]

The advantage of that mathematical form is that resistance does not vanish as the strain rate goes towards zero. The power law also requires the strain rate to vary by orders of magnitude to provide tangible increases in both the modulus and the ultimate strength. The J coefficient and n exponent depend on the nature of the soil. Based on pile driving data, n=0.2 and J=0.3 s⁻⁰·² have been suggested for plastic soils. J should therefore essentially depend on the plasticity of the soil and become quite limited for granular materials.

### 3.6 Degradation Law

When subjected to undrained cyclic loading involving a number N of large strain cycles, the soil structure continuously deteriorates, the pore pressure increases, and the secant shear modulus decreases with N. This process known as *cyclic stiffness degradation* can be best characterized on the basis of strain controlled tests for the type of loading involved with the vibratory penetration of piles. Typical results of strain-controlled tests are sketched in Fig. 5, where the degradation is clearly expressed by the decrease of the amplitude of the peak stress mobilized at successive cycles.

The quantification of the degradation process calls for the introduction of the degradation index $\Delta$, defined by:

$$\tau_n = \Delta \cdot \tau_1 \qquad (9)$$

Laboratory results conducted at constant cyclic strain show that in many soils, the degradation index after N cycles can be approximated by the following relationship as suggested by Idriss et al (1978):

$$\Delta = N^{-t} \qquad (10)$$

The exponent t, called degradation parameter, depends mainly on the amplitude of the cyclic strain and the nature of the material (PI), as suggested by Dobry and Vucetic (1988) and as indicated in Fig. 6 Vucetic, 1993). It is noteworthy that the degradation parameter assumes a zero value at strains smaller than a cyclic "threshold" shear strain, $\gamma_{cv}$ The threshold strain increases with the plasticity of the soil, as suggested in Fig. 6.

### 3.7 Soil liquefaction

Vibration induced compaction of saturated sands has received attention not only from the earthquake engineering community, but also from vibro-compaction specialists.

Recent advances tend to indicate that build up of pore pressures (eventually leading to liquefaction) and volume reduction of cyclically loaded materials are the expression of the same phenomenon, i.e. the irreversible tendency for a particulate arrangement to achieve a denser packing when sheared back and forth.

Under drained conditions, the volume reduction is immediate. Under undrained conditions, the tendency for volume reduction is expressed by an increase in the pore water pressure (see Fig. 7), such that the effective stress is reduced to a value that may be close to zero. It is then necessary to wait for the soil to consolidate in order to see the volume reduction take place.

The strain driven evaluation of the build up of pore pressure as suggested by Dobry et al. (1979) is an approach that lends itself to a direct transposition

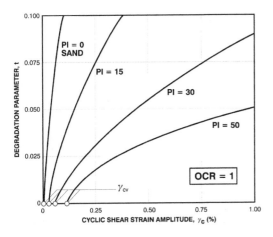

Figure 6. Effect of Plasticity Index (PI) on soil degradation (Vucetic, 1993)

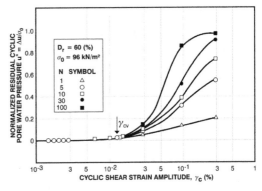

Figure 7. Build up of residual pore pressure in different sands in undrained cyclic strain-controlled tests (Dobry et al., 1982)

to the problem of the vibrations induced by a vertically vibrating pile. It also allows one to evaluate potential changes of the void ratio based on a cyclic strain rather than stress history, as supported by laboratory drained tests conducted on sands by Youd (1972). That framework of analysis entranced by the threshold cyclic strain concept embodies in a single model the intrinsic relationship between degradation and pore pressure build-up, with the advantage that it can be applied to general categories of soils (sands to clays)

The excess pore pressure generated during cyclic loading has been shown (see Fig. 7) to increase with the shear strain and the number of cycles for a given soil type. The damage parameter $\kappa$ approach (Finn, 1981) can be adopted to evaluate the excess pore pressure $\delta u$ resulting from a particular strain history, as characterized by the following equations :

$$du / d\sigma = \lambda / 4 \cdot \ln\left(1 + \kappa/2 \right) \qquad (11)$$

with Relative Energy Loss given by Eq. 5, and
$\kappa = \xi\, e^{\Sigma\gamma}$ (16) with $\Sigma = 5$ and
$\varsigma$ = length of strain path  (12)
    = 4 N $\gamma_c$, for constant amplitude cycles

## 4 PILE VIBRATORY DRIVING TESTING

The above discussion of soil behavior under cyclic loading does not encompass the particular geometry of the profile-soil interface, nor does it consider the continuous penetration of the profile that leads to successive exploration stages into "virgin" soil behavior. That is why a number of experiences have been conducted to reveal soil-structure interaction within a vibratory framework. Based on the ambition and complexity of the tested interface, one can categorize various experiences reported in the literature as conceptual, interface, and both reduced and full-scale testing.

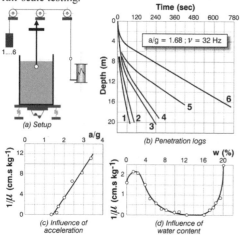

Figure 8. Sphere penetration experiments (after Barkan, 1963)

### 4.1 Conceptual model testing

Tests have been conducted by several Russian researchers to investigate the "vibro-viscous" resistance of soils. In particular, Barkan (1963) reports on the sphere test, shown in Fig. 8a, where a steel ball is sunk into a vibrated soil vessel with the assistance of a bias force. Penetration speed is shown to obey Stokes sedimentation law (see Fig. 8b), allowing one to determine an equivalent viscosity $\mu$. The inverse of that equivalent kinematic viscosity [cm.s/kg] was shown to vary linearly with the relative level of acceleration (a/g), passed a threshold value of approximately 1.4 for a dry sand (see Fig. 8c). The influence of the water content on the "vibro-viscosity factor $1/\mu$" of a sand vibrated at constant a/g is also shown in Fig. 8d, highlighting the near total loss of vibro-penetrability at optimal water content.

### 4.2 Pile-soil interface testing

Soil shear strength resisting the pulling out of a vibrating steel plate against a normal stress controlled medium sand (vibratory direct shear box) has been investigated in the early days by Levchinsky and Savtchencko (Barkan, 1963). The friction coefficient (tan $\phi = \tau/\sigma$) was shown to decrease with cyclic amplitude and frequency. The ultimate relative reduction of the friction was also shown to increase with the grain size within the investigated range shown in Fig. 9. Fig 9 shows that the sand vibratory friction angle can easily drop to ½ to 1/5 of its static value.

### 4.3 Reduced scale tests

Testing of model profiles in soil tanks were initially attempted by Bernhard (1968), Schmid and Hill (1966), continued by Rodger and Littlejohn (1980), Billet and Siffert (1985) and O'Neill et al (1990),

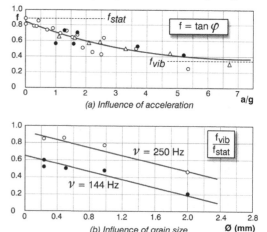

Figure 9. Vibratory friction of sand (after levehinsky and Savtchencko, as reported by Barkan, 1963)

and more recently by Viking (1998) and Holeyman et al (1999). The tests were generally conducted using a lightweight vibrator acting on a heavily instrumented profile. Monitoring included strain gauges, accelerometers and displacement transducers. The soil used was exclusively sand (dry, moist, or saturated), placed at a controlled density, and in some cases, confined at a controlled radial stress. Monitoring of the soil response involved accelerometers, total stress and pore pressure cells during installation as well as compaction and in situ testing after installation.

Insightful observations relative to the vibratory toe resistance have been reported by Schmid (1966), who identified three regimes, depending on the magnitude of the driving force :
- a sinusoidal resistance domain, for a driving force lower than the "resistance threshold"
- an impact domain, when the upward force exceeds the soil uplift resistance; the toe of the pile alternately separates from the soil and tamps it
- a phase instability domain, when the downward force exceeds the soil compressive resistance.

Rodger and Littlejohn (1980) call upon acceleration amplitude to distinguish:
- the elastic state (a<0.6g)
- the trans-threshold state (0.6g<a<1.5g), wherein most of the shear strength reduction takes place,
- the fluidized response state (a>1.5g).

Although their views are contradicted by some of Barkan's observations, these three different states are stated to be confirmed by dynamic direct shear tests performed by others.

Results of tank experiments have been reported in terms of friction reduction coefficients, potential optimal operation, and have shed some light on fundamental soil behavior. Correlations have been established between penetration speed and parameters induced by the vibrator (amplitude, frequency) and by the soil (grain size, relative density, and lateral stress). Although conclusions of the tests conducted under different conditions do not consistently agree, those experiments *generally* identified that:
- penetration speed increased when the relative density decreased and the bias mass increased
- friction was reduced to 30 to 50% of its static value, while a more limited reduction was noted for the toe resistance
- optimum operation of the hammer required at times that the frequency or eccentric moment be reduced, while energy transfer was of the order of 40% of the full theoretical power produced by the vibrator
- a number of observations cannot be explained.

Although reduced scale models are of use, they suffer from improper boundary conditions (at the tank limits) that significantly prevent the vibration energy from propagating away from its source.

### 4.4 *Full scale tests*

Because of inconsistencies in the conclusions derived from reduced scale tests, research has been conducted in several countries based on full-scale tests. Early full-scale programmes have been conducted by Barkan (1963) and Davisson (1970). Other programmes have been conducted by manufacturers on specific equipment, but lead to a limited diffusion of their conclusions. More recently, collective European programs have provided actual penetration speed, but within soil conditions that cannot be controlled, only characterized. Monitoring nowadays involvies acceleration, strain, pore pressure, penetration speed, making the tested profile a fully instrumented probe. Such programs have produced results that have not been fully analyzed or validated (BBRI, 1994, Sipdis, 1997, Viking, 2002); others are being presently conducted (IREX, 2000, Borel et al., 2002). Publication of such research results is usually appreciated by the profession.

## 5 PILE-SOIL-VIBRATOR INTERACTION MODELS

### 5.1 *Types of models*

Models that have been suggested by various authors differ in the way they account for mechanical engineering principles. We will review models purely based on (1) force equilibrium, (2) momentum conservation, (3) energy conservation, and (4) integration of the laws of motion.

### 5.2 *Force equilibrium models*

The force models aim at predicting whether a vibrator can or cannot overcome an estimated soil resistance. They will not provide an estimate of the driving speed. Jonker (1987) and Warrington (1989) have suggested, respectively:

$$F_C + F_i + F_S > \beta_0 \cdot R_{so} + \beta_i \cdot R_{si} + \beta_t \cdot R_t \qquad (13a)$$

$$F_v > \tau_s \cdot \chi \cdot z \qquad \text{provided } s > 2.38 \text{ mm} \qquad (13b)$$

With:
$F_C$ = force generated by the vibrator, per eq. (1)
$F_1$ = inertia forces of dynamic mass, $= M_{dyn} \cdot a$
$F_S$ = surcharge force, per eq. (3)
$\beta_0$ = empirical factor of shaft resistance outside pipe pile,
$R_{S0}$ = soil resistance outside pile shaft,
$\beta_i$ = empirical factor of shaft resistance inside pipe pile,
$R_{si}$ = soil resistance inside pile shaft,
$R_t$ = soil resistance at pile toe.

For sheet-piles Tunker Company recommends to replace $\chi \cdot z$ [m$^2$] with 2.81 times the sheet-pile width.

### 5.3 Energy based models

Energy based models assume the following general form:

$$R.v_p = \beta_t . W_t + (F_i + F_s).v_p \tag{14a}$$

leading to a direct estimate of the penetration speed :

$$v_p = \beta_t \cdot W_t / (R - F_i - F_s) \tag{14b}$$

With:
R   = soil resistance,
$v_p$   = average rate of penetration in m/s,
$W_t$   = theoretical power delivered to the system,
F$i$   = inertia forces of dynamic masses.

Davisson's formula (1970) to estimate the bearing capacity for the Bodine Resonant Driver suggests :

$$\beta_t = 1 - v \cdot s_e \cdot R / 1000 \quad W_t \tag{15}$$

where $s_e$ is an empirically determined set [mm/cycle] representing all energy losses.

Warrington (1989) has coined eq. (14b) as the 'Vibdrive' formula provided a value of 0.1 is used for $\beta_t$ and the power $W_t$ is calculated according to his procedure.

### 5.4 Momentum conservation models

Schmid (1968) has suggested a formula implying that, for steady-state penetration, the momentum of the total mass of the vibrator, additional bias mass ($M_s$), and pile accrued by gravity over a vibration cycle be balanced by the soil resistance impulse:

$$(M_s + M_{vib} + M_p)g \cdot T = \int_0^{T_c} Rdt = \alpha R T_c \tag{16a}$$

with $T_c$ = contact time between pile toe and soil within a cycle and $\alpha$ = coefficient between 0.5 and 1, generally assumed to be 2/3.

Conversely, the penetration speed follows a linear trend passed the threshold acceleration $a_{min}$, which becomes a key parameter to successfully apply the method and estimate $T_c$ :

$$V_p = \frac{(a - a_{min})}{2v} \cdot \left[ \frac{(M_s + M_{vib} + M_p)g}{R / \alpha} \right]^2 \tag{16b}$$

### 5.5 Integration of laws of motion

Comprehensive accounting of the laws of mechanics requires that movement be described at all times from inertial equilibrium conditions. The simplest models involve a single degree of freedom. 1-D models already offer more detailed description of some form of wave propagation, whereas 2-D models might provide future solutions that integrate all types of wave propagation (compression, shear, Rayleigh, etc..).

### 5.5.1 Single degree of freedom (SDOF)

Simplest models of the vibrator suggest that the dynamic mass be the focus of attention, thereby assuming that the pile behaves as a rigid body. Newton's second law can therefore be applied to the dynamic mass :

$$a = \frac{me \cdot \omega^2 \sin(\omega t)}{M_{dyn}} \tag{17a}$$

where

$$M_{dyn} = M_{eb} + M_{cl} + M_p \tag{17b}$$

Holeyman (1993) has suggested a method that integrates the inertial effects of the excess force. That excess force is defined as the difference between the sinusoidal driving force and the opposing soil resistance. A distinction is made between the skin friction, which is reversible (Uplift resistance = Downward resistance) and the toe resistance, which cannot produce uplift resistance. Attention is also paid to the clutch resistance, which is combined with the skin friction.

The soil degraded resistance at the toe and along the shaft is estimated from CPT test results where the friction ratio and acceleration ratio are used to assess the severity of degradation. The method involves an iterative procedure to identify the coexisting acceleration and soil resistance (17b). The driving speed is obtained by intuitively integrating the net downward and upward accelerations over a complete cycle. The method have been verified and liquefaction parameters further refined through calibration with full-scale tests (BBRI, 1994)

Gonin (1998) has followed a similar approach that analytically integrates the effects of an excess force, as shown in Fig. 10. The integration is however performed solely on the toe resistance, while the skin friction influence is accounted for in terms of damping of the driving force. In addition, the wave equation theory is used to estimate the displacement accrued at the toe over the period of net force exceedance.

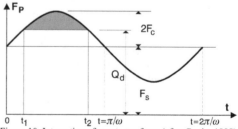

Figure 10. Integration of excess toe force (after Gonin, 1998)

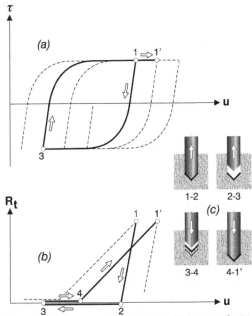

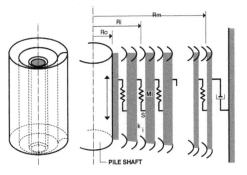

Figure 11. Resistance mobilization versus displacement for (a) skin (b) toe compression (After Dierssen, 1994)

Dierssen (1994) has used a numerical integration scheme to closely follow the time dependence of the skin and toe resistances. Figure 11 provides the shape of the resistance mobilization versus displacement for both skin and toe resistance. One can note that separation of the pile from the soil at the toe is explicitly accounted for.

### 5.5.2 Radial 1-D model

Holeyman (1993b) have suggested the use of a radial discrete model to calculate the vertical shear waves propagating away from the pile. The model, shown in Fig. 12, consists in a succession of concentric cylinders with a linearly increasing depth. The equations of movement are integrated for each cylinder based on their dynamic shear equilibrium in the vertical direction, in a manner similar to that used by Smith (1960) in the longitudinal direction.

Figure 12. Radial 1-D model (Holeyman, 1993b)

The model allows the constitutive relationships described in Section 3 to be readily deployed. The major advantage of that shear wave propagation model is to closely follow the development of degradation as more cycles are simulated. It can also provide insight into vibration levels in the vicinity of the pile. Both features are illustrated by Fig. 13 which provides the effective particle velocity calculated at several distances away from a profile upon vibrator start up. An apparent resonance is indicated, whereas the model does not include a longitudinal or radial dimension that could explain the frequency at which the peak vibration is noted : why? Simply because the model most probably reproduces two soil-pile interaction vibratory modes: the coupled mode and the uncoupled mode.

In the coupled mode (similar to Schmid's sinusoidal domain), the soil remains in contact with the slowly vibrating profile, and the transfer of energy from the pile to the soil is nearly perfect. As the vibrator linearly accelerates (between 0 and 0.5 seconds), vibration levels tend to increase with the square of time since start up. However, as the soil begins to degrade, its shear modulus decreases and the specific shear impedance reduces, leading to loss in the energy transfer. At that point, the coupling between soil and pile suffers some slippage, and therefore time lag. After a sufficient number of cycles, the soil has significantly degraded, and has (60 seconds ageing skipped in Fig. 13) entered into li-quefaction.

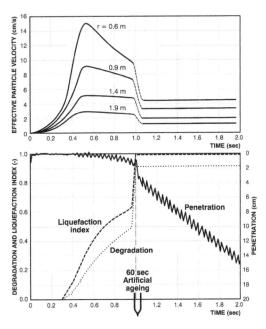

Figure 13. Vibration levels and penetration state parameters estimated upon vibrator startup and regime

13

At the shear modulus of the soil in contact with the profile is nearly zero, and very little energy can pass through the fluidized surrounding zone. The soil in the vicinity of the profile cannot anymore follow the profile movement, from which it uncouples itsel, resulting in a lower level of vibration. That example demonstrates that apparent resonance of soil vibration may be no more than the transient combination of increased rotation speed and soil degradation. The model can also shed light on "damping" as it clearly separates geometric damping from the energy losses attributable to viscous and hysteretic behavior.

A refinement to the above described 1-D radial model was developed by Vanden Berghe and Holeyman (2002). The so-called VIPERE model (for VIbratory PEnetration Resistance), actually implements hypoplastic constitutive behaviour into the geometric model suggested by Holeyman (1994). The soil behaviour is assumed to be hypoplastic and modeled using the Bauer (1996) and Gudehus (1996) constitutive law. More details are provided in Vanden Berghe and Holeyman (2002). One rare feature of the model is its ability to follow pore pressure variations through the various states experienced by the soil during the cycles as a result of dilatant or contractive behaviour of the soil skeleton. Typical simulations for a pile driving problem are presented in Fig. 14 and 15, which emphasize that the three-dimensional character of the volume change trends can be accomodated respectively in a pure shear for the skin friction along the shaft as well as in compression under the pile toe.

### 5.5.3 *Longitudinal 1-D models*

Few authors have adapted Smith's (1960) classic lumped parameters model to represent the longitudinal behavior of a pile subjected to vibratory driving. Gardner (1981) and Chua et al. (1981) have developed a wave-equation computer code where the vibrator is represented by a two-mass system, separated by a soft spring, while the exitor black is subjected to a sinusoïdal force (cfr. eq. (1)); as shown in Fig. 14. The soil behavior is represented by spring-slider-dasplot systems, according to Smith's early suggestion.

Middendorp and Jonker (1988), as well as Ligterink et al. (1990) used the TNOWAVE computer program to analyze the driveability of offshore vibratory driven pipe piles, based on the methods of characteristics. The authors identify the need for a soil model able to describe the degradation of the soil resistance as a function of the oscillation history, and warn that soil parameters may depend on opera- ting frequency and pile movement amplitude.

Moulai-Khatir et al. (1994) have developed together with the University of Houston, the so-called VPDA computer program (for Vibratory Pile Driving Analysis) wherein the action of the hammer is replaced by a static surcharge load and a sinusoïdal load. The soil model was modified from Smith's original in that hyperbolic mobilization curves were adapted for the shaft and toe resistance, as shown in Fig. 17. A simple viscous damper was used to model damping along the shaft, while no damping was deemed necessary at the pile toe.

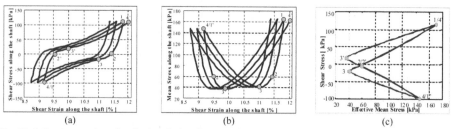

Fig. 14: Soil resistance along the pile shaft during vibratory driving: (a) hysteresis loops, (b) mean stress evolution and (c) vertical shear stress vs effective radial normal stress

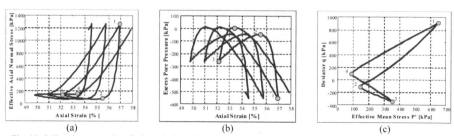

Fig. 15: Soil resistance at the pile base during vibratory driving: (a) evolution of the effective normal axial stress, (b) evolution of the pore pressure, (c) stress path.

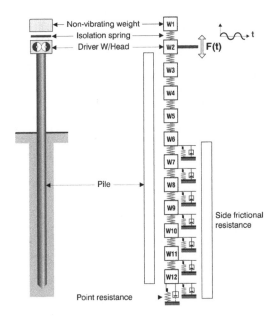

Figure 16. Longitudinal 1-D model

It should also be noted that the GRLWEAP program has included in its latest versions (GRL, 1998) the capacibilty to model vibratory hammers.

## 6 VIBRO-DRIVABILITY ANALYSIS

Most of the models discussed in the previous sections should be able to provide a reasonnable match of calculations with relevant field observations provided the models parameters are properly calibrated.

The use of energy balance methods is discouraged by the author, while force equilibrium methods are of limited use because they do not provide vibratory penetration speed. Momentum based methods may produce a penetration speed very similar to that obtained through integration of the laws of notion of a rigid body. Finally, wave equations methods should not produce penetration speeds significantly different form those obtained form a rigid body analysis, provided the vibrator speed is lower than the resonant frequency of the pile, which is generally the case.
Exceptions to that general case include the Bodine Resonant driver and very long piles for offshore applications (L>50 m).

In the author's opinion, the most critical parameter to assess in order to produce a resonable prediction of vibro-drivability is the soil resistance to vibratory driving.

That is unfortunately where pertinent information and recent consistent experimental data is cruelly missing. The reliability of the predicted vibro-penetration log will strongly depend on the degradation parameters adopted to assess the vibratory penetration resistance form the soil investigation results. The author's experience leads him to use the following crude ultimate degradations coefficients : 0.15 in sand, 0.4 in silt, and 0.65 in clay for skin friction; as well as 0.55 in sand, 0.7 in silt and 0.85 for end bearing.

A more involved assessment of the degradation coefficient has been suggested (Holeyman, 1996) based on CPT test results. In that method, the soil driving resistance is obtained by interpolation between a static value and an ultimately degraded value. The static base ($q_s$) and shaft ($\tau_s$) resistance profiles derived from Cone Penetration (CPT) tests results, i.e. from the cone resistance $q_c$ and local unit skin friction $f_s$ (E1 cone).

The ultimately liquefied base ($q_l$) and shaft ($\tau_l$) unit soil resistances are derived based on an exponential law as expressed below :

$$q_l = q_s \left[ (1 - 1/\Lambda) \cdot e^{-1/FR} + 1/\Lambda \right] \qquad (18a)$$

$$\tau_l = \tau_s \left[ (1 - 1/\Lambda) \cdot e^{-1/FR} + 1/\Lambda \right] \qquad (18b)$$

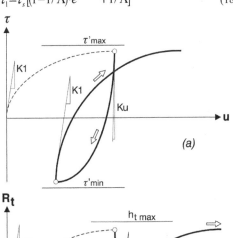

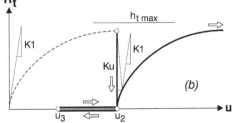

Figure 17. Resistance mobilization for (a) skin friction (b) toe compression (After Moulai – Khatir et al., 1994)

where

$q_l$ = liquefied soil base resistance [kPa]
$\tau_l$ = liquefied soil shaft resistance [kPa]
FR = friction ratio as measured in a CPT test with E1 cone (percentage of the mantle friction to the cone resistance, i.e. FR = 100 $f_s$ /$q_c$)
$\Lambda$ = empirical liquefaction factor expressing the loss of resistance attributable to liquefaction ($\Lambda$ will be higher for saturated and loose sands and is chosen in the range of 4 to 10)

The driving base ($q_d$) and shaft ($\tau_d$) unit resistances are derived from the static and the "liquefied" soil resistance depending on the vibration amplitude following an exponential law as expressed below :

$$q_d = (q_s - q_l)e^{-\alpha} + q_l \qquad (19a)$$

$$\tau_d = (\tau_s - \tau_l)e^{-\alpha} + \tau_l \qquad (19b)$$

where

$q_d$ = driving base unit resistance
$\tau_d$ = driving shaft unit resistance
$\alpha$ = acceleration ratio (= a/g) of the pile, as obtained from Eq. (17a)

At each depth z the vibratory pile driving resistance is calculated :

$$R_{base} = q_d \cdot \Omega \qquad (20a)$$

$$R_{shaft} = \chi \cdot \int_{z=0}^{z=D} \tau_d \cdot dz \qquad (20b)$$

where $\Omega$ is the pile section, $\chi$ the pile perimeter and D the pile penetration.

## 7 BEARING CAPACITY FROM INSTALLATION MONITORING

Because soil resistance degradation is significant during vibratory driving, one should expect it a challenge to estimate the static bearing capacity from the end of penetration vibratory performance of a driven profile.

In the impact driving practice, it is recognized that end of driving (EOD) data generally provides a safe estimate of the pile capacity; that is why beginning of restrike (BOR) or "retap" data is strongly advised to the owner who wishes to tap the value of letting the soil set up. If the end of Vibratory driving (EOV) data is used, methods to estimate the static capacity should allow for recovery of soil degradation, as highlighted in Fig. 2. However significant uncertainty should be expected in the process because the inverse of observed degradation coefficients may range between 2 and 10

That is why extreme caution is warranted when applying so-called pile Vibratory driving (PVD), formulae, even more so than already much detracted (impact) pile driving formulae. A limited number of such PVD formulae have been published; however only one has been, to the author's knowledge been extensively field tested. The "Snipe" formula is therefore the only one that will be discussed in this paper.

The formula is a field-based method was developed in the former Soviet Union according to Steffanof and Boshinov (1977). The following empirical formula is used to predict the static bearing capacity $Q_u$ :

$$Q_u = \frac{1}{\beta}\left(\frac{25.5 \cdot W}{v \cdot s} + F_t\right) \qquad (21)$$

where

$Q_u$ = load capacity, in [kN];
W = power used by the vibrator to drive the pile, in [kW]
$F_t$ = total weight (force) of vibro-hammer and pile, in [kN]; = ($M_{vib}$ + $M_p$)g
$1/\beta$ = empirical loss coefficient (in Soviet practice $1/\beta$ is safety taken to be 5 in cohesionless soils) reflecting the influence of driving on soil properties.

The bearing capacity in of vibro-driven pipe piles has been verified by PDA monitoring the behavior of the finished product at the beginning of impact restrike. This of course requires that a specific BOR procedure be enforced a sufficient time after the EOV installation, in order to allow pore pressures to dissipate and accrue soil setup.

The monitoring of the installation of vibratory driven piles is not at all as widely spread as for impact driven piles. Recent improvements in the field monitoring devices (PDA, TNO-System, etc) now allows the geotechnical engineer to control pile acceleration, stress and energy. However, there is no equivalent interpretation of the data to the CASE formula or CAPWAP method, available for nearly 15 years in the impact driven products.

Field monitoring provides a tremendous advantage in controlling the effective performance of the vibratory hammer, as illustrated by the following case history reported by Holeyman et al. (1996).

That case involved the installation of a 20.6 m long tubular steel pile with a thickness of 9.5 mm and a diameter of 1 m on a site in Kortrijk (B). Figure 18 shows the subsoil profile as depicted by a CPT test (cone M4) performed at the site. The water table was encountered at a depth of −1.8 m. The upper twelve meters consist of very soft river deposits; below the underlying sandy layer was found a very stiff tertiary clay layer in which the tube had to be driven.

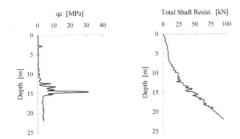

Figure 18. Subsoil profile site at Kortrijk

A preliminary calculation using eqs. (17) through (20) pointed out that the necessary time to install the pile to a depth of 20 m with a PTC

30HFV vibratory hammer was 13½ minutes. However, driving met refusal at a depth of 11 m.

The reason for the difficult driving and the difference between the predicted and the observed penetration speed was explained by measurements taken during the actual driving of the pile. The pile vibration amplitude was measured by means of a velocity transducer placed at the pile head and a velocity transducer (protected by a cover) at the pile toe.

Figure 19 shows the monitored amplitude of vibration at the pile top and at the pile base upon loss of drivability. The observed frequency was 38 Hz.

From the measurement results, one can observe that :

- the vibration amplitude at the pile top (0.65 mm zero to peak) is considerably less than the nominal vibration amplitude which is,

$$\frac{me}{M} = \frac{me}{M_{vibr} + M} = \frac{26000 kg.mm}{(6500 + 4820)kg} = 2.3 mm$$

- the amplitude at the pile base (0.45 mm zero to peak) is smaller than the amplitude at the pile top (0.65 mm)

It would appear that the pile base amplitude (0.45 mm) is not sufficient to allow the pile to penetrate as the stress-strain behaviour for clayey soils is primarily elastic for small amplitudes. Possible explanations for that observation were considered :

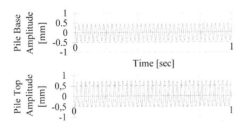

Figure 19. Record from the vibration amplitude upon refusal

- An important soil (i.e. clay) mass was sticking to the vibrating pile, leading to a more important vibrating mass, leading to a smaller vibration amplitude.
- The vibratory hammer was unable to deliver the required energy, and thus maintain its nominal amplitude or frequency. A characteristic of the PTC variable eccentric hammers is that a lack of power results in a reduction of vibration amplitude (rather than a reduction of frequency (Houzé, 1994)).
- A smaller amplitude at the pile base was obtained due to the elasticity of the pile.

By applying the observed vibration amplitude to the calculation model (Figure 20), a much better correlation between the calculated and the observed penetration time was obtained. The pile was placed at the bottom of an excavation at –2.5 m and penetrated 4.5m under its own weight. As a result, observed and calculated penetration rates are reported starting at level - 7 m. Figure 20 evidences that the difference for the predicted and observed penetration times for the site in Kortrijk was not due to an incorrect estimation of the dynamic soil resistance but due to an incorrect estimation of the vibration amplitude, which happened to be limited by the nominal power of the power pack. A more powerful power pack was brought on site and the piles could be vibrated to design depth using the same vibrator.

## 8 SUGGESTIONS FOR FURTHER RESEARCH, DESIGN, AND PRACTICE

After reviewing the present state-of-the-art of vibratory driving, the following suggestions for further consideration are offered :

- soil mechanics research is needed in the area of large cyclic deformation to better understand and assess the effects of degradation and liquefaction under those extreme conditions,

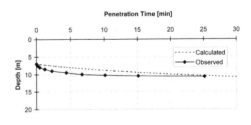

Figure 20. Predicted and Observed penetration log at Kortrijk site compared with predicted log using actual vibratory amplitude

17

- full scale vibratory driving tests, with extensive field monitoring, will be required preferably to reduced scale laboratory tests, which suffer from improper energy dissipation boundary conditions,
- potential and transferred power of vibrators need to be better defined, as well as modeled for better description of the mechanical behavior of vibrators,
- peak vibration of the soil surrounding a profile upon vibrator start up does not necessarily imply soil resonance; it can also result from the combination of increasing frequency and degrading soil resistance,
- monitoring of vibrated profiles is recommended with the view to emulate the benefits accrued by a similar practice for driven profiles, and
- procedures for vibratory loading tests should be developped.

## 9 REFERENCES

Bauer, E. (1996). "Calibration of a Comprehensive Hypoplastic Model for Granular Materials." Soils and Foundations, Vol. 36, N°1, pp.13-26.

Barkan, D.D., (1963), Méthodes de vibration dans la construction, Dunod, Paris, 302 p. *(French translation of Original in Russian " Vibrometod V Stroïteljstve, 1960)*

BBRI .(1994). High performance vibratory pile drivers base on novel electromagnetic actuation systems and improved understanding of soil dynamics, *Progress reports of the BRITE/EURAM research contract CT91-0561, 1994.*

Bernhard, R.K., (1967), Fluidization phenomena in soils during vibro-compaction and vibro-pile-driving and –pulling. Hanover, NH 1967, 58 pp., *US Army Cold Regions Research and Engineering.*

Billet, P., Siffert, J.G., (1989). "Soil-sheet pile interaction in vibro-piling" *Journal of Geotechnical Engineering, ASCE*, Vol. 115, N° 8, pp. 1085-1101

Borel, S. et al. (2002), "Full-scale behaviour of vibratory driven piles in Montoir", Proceedings of Transvib2002 Conference, Louvain-la-Neuve, Sept 9-10, 17 p.

Chua, K.M., Gardner, S., Lowery, L.L., (1987). "Wave Equation Analysis of a Vibratory Hammer-Driven Pile", *Proc. Offshore Technology Conf.*, Vol. 4, pp. 339-345

Davisson, M.T., (1970). "BRD Vibratory driving formula", *Foundation facts*, Vol. 1, N° 1, pp. 9-11

Dierssen, Guillermo, (1994), "Ein Bodenmechanisches Modell zur Beschreibung des Vibrationsrammens in körningen Böden", *Doctoral Thesis, University of Karlsruhe, Germany*

Dobry, R., Ladd, R.S., Yokel, F.Y., Chung, R.M. and Powell, D. (1982). "Prediction of Pore Water Pressure Buildup and Liquefaction of Sands During Earthquakes by the Cyclic Strain Method". *National Bureau of Standards Building Science Series 138*, July 1982, 150 pp.

Dobry, R. and Swiger, W.F. (1979). "Threshold Strain and Cyclic Behavior of Cohesionless Soils". *Proc. 3rd ASCE/EMDE Specialty Conference.* Austin, Texas, pp. 521-525;

Dobry, R. and Vucetic, M. (1987). "State-of-the-Art Report: Dynamic Properties and Response of Soft Clay Deposits". *Proceedings of the Intl. Symposium on Geotechnical Eng. of Soft Soils*, Mexico City, Vol. 2, pp. 51-87.

Drnevich, V.P., Hall, J.R., Jr., and Richart, F.E., Jr. (1967). "Effects of Amplitude of Vibration on the Shear Modulus of Sand." *Proceedings of the International Symposium on Wave Propagation and Dynamic Properties of Earth Materials*, Albuquerque, N.M., pp. 189-199.

Finn, W. D. L. (1981) "Liquefaction Potential: Developments Since 1976", *Proceedings, Intl. Conf. on Recent Advances in Geotechnical Earthquake Engineering and soil Dynamics, St. Louis*, Missouri, Vol. II, pp. 655-681.

Gardner, Sherrill, (1987). "Analysis of vibratory driven pile". *Proc. of 2nd Int. Conf. on Deep Foundation*, Luxembourg, 5-7 May, pp. 29-56.

Gonin, J. (1998), Quelques réflexions sur le vibrofonçage, *Revue Française de Géotechnique*, N° 83, 2ème trimestre, pp. 35-39

Gudehus, G. (1996). "A comprehensive constitutive equation for granular materials. " Soils and Foundations, Vol. 36, N° 1, 1-12.

Hardin, B.O. and Black, W.L. (1968). "Vibration Modulus of Normally Consolidated Clay." *Journal of the Soil Mechanics and Foundations Division, ASCE*, Vol. 94, No. SM2, Proc. Paper 5833, pp. 353-369.

Holeyman, A. (1985) "Dynamic non-linear skin friction of piles," *Proceedings of the International Symposium on Penetrability and Drivability of Piles*, San Francisco, 10 August 1985, Vol. 1, pp. 173-176.

Holeyman, A. (1988) "Modeling of Pile Dynamic Behavior at the Pile Base during Driving," *Proceedings of the 3rd International Conference on the Application of Stress-Wave Theory to Piles*, Ottawa, May 1988, pp. 174-185.

Holeyman, A. (1993a) "HYPERVIB1, An analytical model-based computer program to evaluate the penetration speed of vibratory driven sheet Piles", Research report prepared for BBRI, June, 23p.

Holeyman, A. (1993b) "HYPERVIBIIa, An detailed numerical model proposed for Future Computer Implementation to evaluate the penetration speed of vibratory driven sheet Piles", Research report prepared for BBRI, September, 54p.

Holeyman, A. & Legrand, C. (1994). Soil Modelling for pile vibratory driving, International Conference on Design and Construction of Deep Foundations, Vol. 2, pp 1165-1178, Orlando, U.S.A., 1994.

Holeyman, A., Legrand, C., and Van Rompaey, D., (1996). A Method to predict the driveability of vibratory driven piles, *Proceedings of the 3rd International Conference on the Application of Stress-Wave Theory to Piles*, pp 1101-1112, Orlando, U.S.A., 1996.

Houzé, C. (1994). HFV Amplitude control vibratory hammers : piling efficiency without the vibration inconvenience, in DFI 94, pp. 2.4.1 to 2.4.10, *Proceedings of the Fifth International Conference and Exhibition on Piling and Deep Foundations*, Bruges, Belgium, 1994.

Idriss, I.M., Dobry, R, and Singh. R.D. (1978). "Nonlinear Behavior of Soft Clays during Cyclic Loading." *J. Geotechnical Engineering Div.*, ASCE, 104(GT12), pp. 1427-1447.

IREX (1998), Vibrofonçage des pieux et palplanches – Etude exploratoire, (by Le Tirant, P., Borel, S., Gonin, H., Guillaume, D. and Longueval, A.)

Jonker, G., (1987). "Vibratory Pile Driving Hammers for Oil Installation and Soil Improvement Projects". *Proc. of Nineteenth Annual Offshore Technology Conf.,* Dallas, Texas, OTC 5422, pp. 549-560.

Kondner, R. L., (1963). "Hyperbolic Stress-Strain Response: Cohesive Soils." *Journal of the Soil Mechanics and Foundations Division,* ASCE, Vol. 89, No. SM1, pp. 115-143, Jan.

Ligterink, A., van Zandwijk, C., Middentorp, P., (1990). "Accurate vertical pile installation by using a hydraulic vibratory hammer on the Arboath project". *Proc. 22nd annual Technology Conf. in Houston*, Texas, May 7-10, pp. 315-326.

Masing, G. (1926), "Eigenspannungen und Verfeistigung beim Messing", *Proceedings of Second International Congress of Applied Mechanics*, pp. 332-335.

Midendorp, P. and Jonker, G. (1988), Prediction of Vibratory Hammer Performance by Stress wave Analysis, *Preprint to the 3rd Int. Conf. on the Application of Stresswavre Theory to Piles,* Ottawa

Moulai-Khatir, Reda. O'Neill, Michael W., Vipulanandan, C., (1994). "Program VPDA Wave Equation Analysis for Vibratory Driving of Piles", *Report to The U.S.A. Army Corps of Engineers Waterways Experiments Station. Dept. of Civil and Environmental Engineering,* UHCE 94-1, Univ. of Houston, Texas, August 1994, 187 pp.

Novak, M., Nogami, T., and Aboul-Ella, F. (1978). "Dynamic Soil Reactions for Plane Strain Case", *J. Engrg. Mech. Div.,* ASCE, 104(4), 953-959.

NRC (1985). "Liquefaction of Soils During Earthquakes." National Research Council Committee on Earthquake Engineering, Report No. CETS-EE-001, Washington, D.C.

O'Neill Michael W., Vipulanandan, C., (1989). "Laboratory evaluation of piles installed with vibratory drivers". *National Cooperative Highway Research Program,* Report n° 316, National Research Council, Washington, DC. Vol. 1 pp. 1-51. ISBN 0-309-04613-0

O'Neill Michael W., Vipulanandan, C., Wong, D., (1990). "Laboratory modelling of vibro-driven piles". *Journal of Geotechnical Engineering, ASCE,* Vol. 116, N° 8, pp. 1190-1209

Robertson, P.K. and Wride, C.E. (1998). "Evaluating cyclic liquefaction potential using the cone penetration test", Canadian Geotechnical Journal, No. 35, pp. 442-459.

Rodger, A.A. and Littlejohn, G.S. (1980). "A study of vibratory driving in granular soils". *Geotechnique,* Vol. 30, n°

3, pp. 269-293.

Seed, H.B. and Idriss, I.M. (1970). "Soil Moduli and Damping Factors for Dynamic Response Analyses." Earthquake Engineering Research Center, College of Engineering, University of California, Berkeley, Report No. EERC 70-10.

Seed, H.B. and De Alba, P. (1986) "Use of SPT and CPT Tests for Evaluating the Liquefaction Resistance of Sands" *Proc. INSITU 86*, VA, 22 p.

Schmid, W.E., (1969). "Driving resistance and bearing capacity of vibrodriven model piles". *American Society of Testing and Materials Special Techn.,* Publ. 444, pp. 362-375

Schmid, W.E. & Hill, H.T. (1967). "A rational dynamic equation for vibro driven piles in sand". *Symp. Dynamic Properties of Earth Materials. New Mexico University*, p. 349.

Smith E.A.L., (1960). "Pile-driving analysis by the wave equation". *Journal of the Soil Mechanics and Foundations Divisions, ASCE,* Vol. 86, August 1960.

Vanden Berghe, J-F, (2001) "Sand Strength degradation within the framework of pile vibratory driving", doctoral thesis, Université catholique de Louvain, Belgium, 360pages.

Viking, K., (1997). "Vibratory driven pile and sheet piles – a literature survey", *Report 3035, Div. of Soil and Rock Mechanics, Royal Institute of Technology,* Sweden, 75 p.

Viking, K., (2002). "Vibratory driveability –a field study of vibratory driven sheet piles in non-cohesive soils" *PhD thesis 1002, Div. of Soil and Rock Mechanics, Royal Institute Technology,* Stockholm, Sweden, 281 pp, ISSN 1650-9501.

Vucetic, M. and Dobry, R. (1988). "Degradation of Marine Clays Under Cyclic Loading. *ASCE Journal of Geotechnical Engineering,*Vol. 114, No.2,pp.133-149.

Vucetic, M. and Dobry, R. (1991). "Effect of Soil Plasticity on Cyclic Response." *ASCE Journal of Geotechnical Engineering, Vol.* 117, No. 1, pp. 89- 107.

Vucetic, M. (1993). "Cyclic Threshold Shear Strains of Sands and Clays", Research Report, UCLA Dept. of Civil Engineering, May 1993.

Vucetic, M. (1994). "Cyclic Threshold Shear Strains of Sands and Clays", Paper in print, *ASCE Journal of Geotechnical Engineering*

Warrington, Don. C., (1989). "Driveability of Piles by vibration". *Deep Foundations Institute 14th Annual members Conf.,* Baltimore, Maryland, USA, pp. 139-154.

Westerberg, E., Massarch, K. Rainer. Eriksson, K., (1995). "Soil resistance during vibratory pile driving", *Proc. to Int. Symposium on Cone Penetration Testing*, Linköping, Sweden, Vol. 3, Report 3.41, pp. 241-250.

Youd, L.T. (1972). Compaction of sands by repeated shear straining, *Journal of the Soil Mechanics and Foundations Division*, 1972, Proc. ASCE, Vol. 98, SM7, pp. 709-725.

*Vibratory Pile Driving and Deep Soil Compaction - TRANSVIB2002,*
*Holeyman, VandenBerghe & Charue (eds.), © 2002 Swets & Zeitlinger, Lisse, ISBN 90 5809 521 5*

# Modeling of vibratory pile driving

F.Rausche
*GRL Engineers, Inc., Cleveland, Ohio, USA*

ABSTRACT: Vibratory installation of piles and casings can be extremely economical and therefore the contractors' preferred method of pile driving. Compared to impact pile driving it also has the advantage of reduced noise pollution. Unfortunately, this installation method is still fraught with uncertainties. The foremost question is driveability and equally important is the question of bearing capacity of the installed pile. Both are difficult to answer prior to actually driving a pile. In other words, given a soil profile and pile type, the question of which vibratory hammer could drive the pile to a certain depth often cannot be answered with sufficient certainty. Once the pile has been installed, its bearing capacity cannot be calculated with sufficient accuracy based on the observed final rate of penetration.

Many attempts have been made to calculate the pile capacity based on the driving resistance, however, to date it is still required that the pile is redriven with an impact hammer for acceptance as a bearing pile. As far as driveability is concerned, simple charts issued by hammer manufacturers based on pile size or weight seem to be as reliable as other more sophisticated methods of hammer selection.

This paper summarizes current analytical methods and explains how wave equation analysis can be used in a manner comparable to the analysis of impact driven piles. Hammer and soil modeling details are discussed. A few examples demonstrate capabilities and limitations of the methods and, for the wave equation approach, sensitivity of results to important soil resistance parameters.

## 1 INTRODUCTION

The history of vibratory pile driving has been discussed by several authors among them Smart (1969), Leonards et al. (1995) and, Viking (2002). The latter dissertation is very much up to date and little can be added to its literature review. Viking reports that the first studies on vibratory pile drivers were done in Germany in 1930 and in Russia in 1931. The first production units were built in Russia during or after the Second World War. Their utility was quickly recognized, leading to new developments in France, Germany and the United States. Among the newer developments was the Bodine hammer, which produced frequencies in excess of 100 Hz. Called the Resonant Pile Driver, this machine achieved much higher penetration rates than those with "normal" frequencies at or below 20 Hz. Although resonance is a function of the mass of the driver and the size of the pile, it is probably reasonable to draw the dividing line between resonant and low frequency pile drivers at 50 Hz.

The analytical treatment of vibratory pile drivers has either been done with simple energy formulas or with discrete representations of pile and/or soil. Discrete models include integration, much as introduced by Holeyman, et al. (1996), or so-called wave equation analyses, e.g., GRLWEAP (GRL, 1998). Additionally finite element analyses have been tried (Leonard, et al. 1995) as an improvement over other methods.

## 2 OBJECTIVES OF VIBRATORY HAMMER MODELING

Vibratory installation of preformed piles and casings has an important economic impact. Installation times, only 10% of those achieved with impact hammers, are not uncommon. On the other hand, unexpected refusal may occur where impact hammers still drive efficiently. Also the vibratory hammer may produce undesirable vibrations in nearby structures, a limitation that will not be discussed in this paper.

The economic advantage of the vibratory hammer can only be realized if the contractor correctly predicts which size hammer will drive the pile to a required depth. Additional savings would be realized if it were assured that the pile had the required bearing capacity after installation. The simulation of the installation by vibratory hammers should therefore enable the analyst to predict the rate of penetration vs. depth (driveability analysis). In addition, a so-called bearing graph should be constructed, which would relate the pile bearing capacity to the rate of penetration at the end of the installation. Examples of these relationships will be demonstrated below.

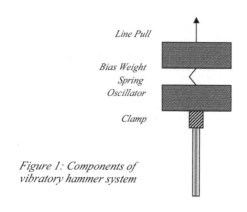

*Figure 1: Components of vibratory hammer system*

## 3 BASIC COMPONENTS OF A VIBRATORY HAMMER

The vibratory hammer, in its most common form, consists of pairs of eccentrically mounted masses which are contained in a frame whose appreciable mass may be called the oscillator. A bias mass isolates the oscillator from the hammer support – usually a crane line. The oscillator is separated from the bias mass by a very soft spring (Figure 1). The bias mass therefore adds a static force to oscillator and pile. The force in the crane line reduces this static force and, if it is greater than all weights, allows for pile extraction. Conveniently, the pile is attached to the oscillator by means of a hydraulic clamp. This connection may be considered rigid and, for modeling purposes, the clamp can be considered an integral part of the oscillator mass.

When eccentrically supported masses (combined eccentric mass $m_e$) spin at a rotational frequency $\omega = 2\Pi\, f$ (f in Hz), their centrifugal force is

$$F_c = m_e\, \omega^2 \qquad (1)$$

This centrifugal force (actually it differs slightly from Eq. 1 because of the oscillator's vertical motion) is transmitted through the eccenter mass bearings to the oscillator and thus to the pile. Only vertical components of the centrifugal force are transmitted to the pile because pairs of eccenters are spinning in opposite directions. Normally the hammer frequency is between 20 and 40 Hz and each peak compressive force generated by the vibratory hammer therefore occurs at intervals of 25 to 50 ms. For a resonant hammer, successive peak force values may occur at intervals of only 8 or 10 ms.

The free-free frequency, $f_F$, of a 20 m steel pile of wave speed c = 5120 m/s is

$$f_f = c/2L = 5120/40 = 128 \text{ Hz}$$

Comparing this free pile frequency with that of a low frequency hammer shows that resonance is unlikely in piles of normally encountered lengths. However, the mass of the vibratory hammer and the clamp, attached through the clamp to the pile, tends to reduce the lowest frequency of the overall system and, when piles get long and drivers heavy, makes resonance possible. According Poulos et al. (1980) the lowest resonance frequency of a system roughly reduces to 50% of the pile frequency if the mass on top of the pile equals the weight of the pile.

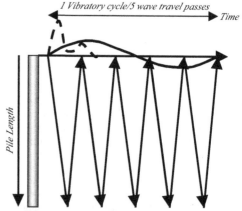

*Figure 2: Comparing typical impact and vibratory hammer records*

If the hammer frequency is significantly lower than the hammer-pile frequency, then the particles of the pile have practically the same direction of motion at the same instance in time. Figure 2 demonstrates in a length-time diagram the different loading patterns of a vibratory hammer and an impact hammer. The relationship shown is approximately scaled for a steel pile of 20 m length and a vibratory hammer with a 25 Hz frequency.

Because of the relatively slow upward and downward motion of the pile under a vibratory hammer, many mathematical models ignore the elasticity of the pile and treat the pile as a single mass, acted upon by the oscillator force. The soil may then be considered an ideally plastic material, which reacts with an upward directed force during the downward pile motion and in downward direction when the pile is pulled upwards.

## 4 RESONANCE EFFECTS

It is important to distinguish two resonance effects:

(1) Resonance in the soil has been observed to occur at relatively low frequencies (e.g. around 10 Hz) therefore occurs primarily during hammer start-up and shutdown, unless the hammer is equipped with a variable eccentric moment that can be reduced before the frequency is varied through the resonance range. This effect has been discussed in detail by Massarsch (1992). Only those models that include the mass and stiffness properties of the soil surrounding the pile have a chance of correctly predicting the soil resonance phenomenon.

(2) Resonance in the hammer/pile system may occur at several frequencies. As discussed earlier, a low frequency resonance is possible depending on the relative magnitude of the hammer and pile masses. Resonance will also occur near the piles basic frequencies. Only those pile models that represent it's flexibility have a chance of correctly predicting hammer/pile resonance.

## 5 REFUSAL CRITERION

Refusal is defined as a certain limiting rate of penetration (mm/s). Smart defines it as 6.2 mm/s. Viking suggests 8 mm/s, citing the danger of excessive heat development in sheet pile locks. Of course, where bearing piles are driven, the lock friction is not of concern. In that case the pile could be driven to lower rates of penetration. For example, impact hammers typically are used to sets as low as 1 mm per blow with blow rates around 1 blow/s. Thus, from a productivity point of view, 1 mm/s still appears to be an acceptable rate of penetration.

Rate of penetration is not necessarily the only criterion for refusal conditions. Stresses in the pile, particularly around the clamp also must be considered. A realistic pile model can be particularly helpful for hammer and pile selection if the hammer is capable of predicting accurate stress levels near hammer/pile resonance.

## 6 SIMPLIFIED APPROACHES AND ENERGY MODELS

Consider the basic energy formula used for impact driven piles

$$R_u = \eta E / (s + s_L) \qquad (3)$$

Where $R_U$ is the ultimate pile capacity; 0 is an efficiency, which reduces the theoretical hammer energy, E, to its actual value; s is the set per blow and $s_L$ is a displacement value covering losses in pile and soil. Adding to the reduced hammer energy the energy that the system's weight, W, (hammer and clamp minus crane line pull) adds, one obtains

$$R_u = (\eta E + Ws)/ (s + s_L).$$

Dividing the numerator and denominator of this equation by the time for one complete cycle $T = 1/f$, then this formula becomes

$$R_u = (P + W \, v_R)/ (v_R + f \, s_L) \qquad (4)$$

Table 1a: *Pile properties after Smart (1969)*

| Pile | Type | Length (est.) | Area | Penetr. (est.) | Hammer Power | Hammer Frequency | Soil |
|------|------|------|------|------|------|------|------|
| | | m | cm$^2$ | m | kW | 1/s | |
| 62,1 | HP14x117 | 30 | 221 | 27 | 343 | 107 | Silt; dense to very dense Sand |
| 62,2 | HP14x117 | 30 | 221 | 27 | 310 | 113 | Silt; dense to very dense Sand |
| 62,3 | HP14x117 | 30 | 221 | 27 | 343 | 107 | Silt; dense to very dense Sand |
| 78, 1 | HP14x117 | 30 | 221 | 18 | 343 | 91 | Cemented Sand; Clay |
| 83, 1 | CE-Pipe | 24 | 47-54 | 20 | 37 | 43 | Sand: N=31 to 61 |
| 83, 2 | CE-Pipe | 24 | 48 | 20 | 37 | 49 | Sand: N=31 to 61 |

where P is the power actually supplied by the vibratory hammer's power unit (for that reason an efficiency factor is not needed), $v_R$ is the rate of penetration (averaged over one cycle) and $s_L$ is a loss term, to be determined empirically as in a pile driving formula. This is the Davisson formula according to Smart (1969). Recommended values for $s_L$ range between 0.001 and 0.1 and may average 0.03. Also, according to Smart, Bernhard performed model pile studies and modified the power formula as follows:

$$R_u = (\lambda P / v_R )(L/D) \qquad (5)$$

The loss factor $\lambda$ not only covers power losses, it is an empirical adjustment factor and is to be set to 0.1 unless other correlation data exists. The L/D (pile length divided by pile penetration) may have an effect when the pile is only partially driven.

Davisson's power formula was modified and tested by Smart on more than 60 cases where rate of penetration and power readings were available. In several cases for which load test information was also available, Smart developed load-set curve dependent adjustment factors in a so-called "permanent-set method". This method is not further discussed here as it adds undue complexity to what should be a simple formula.

The case studies demonstrated by Smart represented primarily piles installed with a Bodine Resonant hammer with frequencies between 43 and 144 Hz and system weight W=98kN. Results from Equation 4 and 5 are demonstrated in Table 1b for 5 of Smart's load test cases, described in Table 1a.

*Table 1b: Results from Power Formulas*

| Pile | Rate of Penetr. | Load Test | Davisson | | | Bern-hard |
|---|---|---|---|---|---|---|
| | | | 0.1 | 0.03 | 0.001 | |
| | mm/s | kN | kN | kN | kN | kN |
| 62,1 | 132.1 | 2314 | 882 | 1667 | 2642 | 289 |
| 62,2 | 8.9 | 2492 | 1049 | 3268 | 26399 | 3869 |
| 62,3 | 15.2 | 3560 | 1201 | 3561 | 19192 | 2502 |
| 78,1 | 4.6 | 2270 | 1458 | 4649 | 49918 | 12509 |
| 83,1 | 91.4 | 490 | 230 | 372 | 500 | 51 |
| 83,2 | 67.6 | 668 | 229 | 419 | 638 | 69 |

These results show that Bernhard's formula yields rather unreliable results with minima and maxima between 10% and 550% of the load test result. Davisson's formula varies between 34 and 2200% if the full recommended range of adjustment factors is considered. The medium factor of $s_L =$ 0.03 inch/s produces a scatter between 63 and 205%. This relatively good result may be attributable, at least in part, to the fact that this same data was included in the study that led to the recommended loss factors. For other sites and hammers, different loss factors may be needed.

The simple power balance equations suffer from a very important defect: they do not consider the relative magnitudes of end bearing and shaft resistance, which, as we shall see affect driveability and bearing capacity evaluations to a significant degree. Furthermore, power formulas do not consider the type of soil into which the piles are driven.

Most disturbing, however, for all potential capacity determination methods is in Smart's data the apparent fact that the rate of penetration is unrelated to bearing capacity; for illustration, Figure 3 depicts load test capacity vs. penetration rate for the four H-piles of Tables 1a and 1b. Note that soil types were similar and that power and frequency varied only slightly for these four cases.

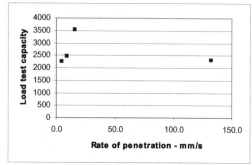

*Figure 3: load test capacity vs. rate of penetration for the four piles of Table 1*

7 CASE METHOD

The Case method was developed in the 1960s for impact driven piles on which measurements of force, F(t), and velocity, v(t), are taken near the top of the pile during driving. The Case Method formula for the evaluation of the instantaneous static resistance force, R(t), was derived assuming an elastic pile and a soil resistance that acts in one direction (upward when the pile moves downward). The Case Method formula then becomes (Rausche et al., 1985)

$$R(t) = \tfrac{1}{2} (F1+Zv1)(1-Jc) + \tfrac{1}{2}( F2-Zv2)(1 + Jc) \quad (6)$$

Where F1 is the measured force at time t
   F2 is the force at time t + 2L/c
   v1 is the velocity at time t
   v2 is the velocity at time t + 2L/c
   Z = EA/c is the pile impedance
   Jc is a dimensionless damping factor
   L is the pile length
   c is the wave speed in the pile material
   E is Young's modulus of the pile material
   A is the cross sectional area of the pile material.

24

For a rigid body the time 2L/c reduces to zero and thus F1 = F2 = F(t) and v1 = v2 = v(t) and the formula becomes

$$R(t) = F(t) + M\,a(t) - J_v\,v(t). \qquad (7)$$

Eq. 7 is adequate for low frequency hammers. For higher frequencies, near the hammer-pile system's resonance level, the Case Method equation for elastic piles (6) would be more reasonable. Formula (6) approaches (7) as the pile length approaches zero and the pile becomes a rigid body of mass, M, with acceleration a(t). The damping factor, $J_v$, is equivalent to the product of Z Jc.

### 7.1 Example – Case Method applied to an offshore pile

For low frequency hammers, Eq. 7 is satisfactory as shown by Likins et al. (1992) who described how a 1520 mm diameter pipe of 40 m length was driven with an ICE 1412 hammer (115 kg m, 21 Hz, 410 kW) through 13.4 m of sand into a firm to very stiff clay. The pile met refusal at a depth of 19.2 m and was then driven with a Vulcan 060 steam hammer starting at a blow count of 75 blows per 0.3 m of penetration. A reasonably accurate correlation was obtained between the positive peak value calculated by the rigid body Case Method formula and a dynamic impact test following the vibratory installation. In this case $J_v$ was set to zero. (The lack of damping reduction may have been offset by the soil setup occurring between end of vibratory driving and the beginning of the impact test.) Figure 4 depicts a portion of the record taken at the end of the vibratory driving together with related, calculated pile variables. The force and acceleration records were evaluated according to Eq. 7 and indicated 2820 kN peak soil resistance while the impact records yielded a CAPWAP capacity of 3050 kN.

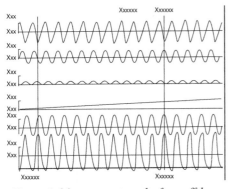

Figure 4: Measurement results from offshore pile, from top: acceleration, velocity, displacement, transferred energy, force and resistance after Eq. 7.

## 8 INTEGRATION METHODS

The name "Integration Methods" was used in Viking (2002) for an approach which (a) formulates a force balance for a rigid pile, (b) uses Newton's Second Law to calculate acceleration and (c) integrates the acceleration to obtain the rate of pile penetration. Viking uses this name for rigid body models of the pile and either concentrated or discretized soil models such as the model of Holeyman et al., 1996. Viking differentiates between pile integration methods having rigid pile models from wave equation type models even though the latter also integrate motions calculated from a force balance. The difference is that the process is repeated for the segments of a discretized elastic pile.

The basic model of Holeyman et al. (1996) is depicted in Figure 5; it not only includes the shaft resistance and end bearing but also lock friction. Actually, shaft resistance and lock friction are treated in

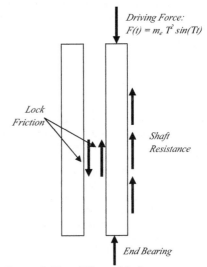

Figure 5: Sheet Pile model after Holeyman et al. 1996.

the same manner, i.e. as ideally plastic resistance components. More importantly, the shaft resistance is distinguished from end bearing by allowing it to have a negative downward resistance during the upward pile motion while end bearing only has a positive component. Holeyman's simplified model calculates the velocities during the upward and downward motions. The resistance components are modeled as ideal plastic forces acting at shaft and toe. An important part of Holeyman's model is the reduction of the static shaft and toe resistance to so-called liquefied values. The algorithm requires an iterative analysis of soil and pile resistance since the liquefaction of the soil is considered a function of the vibration amplitude.

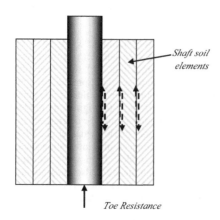

Figure 6: Vipere pile and soil model, after Viking 2002

Viking describes also the more elaborate Vipere soil model after Vanden Berghe (Figure 6). It includes a hyperbolically behaving shaft resistance and a practically bilinear toe resistance. The Vipere shaft resistance calculation is based on the analysis of several cylindrical soil elements, as proposed by Holeyman et al.(1994), surrounding the pile to model radiation damping associated with the shaft resistance. A degradation algorithm of the shaft resistance as a function of soil strain is also included in this model. The shaft resistance model appears to be symmetric, i.e. the upwards directed resistance forces and the downwards directed resistance forces have equal magnitude. The Vipere soil model parameters are based on laboratory test results.

The Vipere toe model is shown in Figure 7. It is interesting to compare this model with the wave equation toe model discussed below and depicted in Figure 8c. Both models move with zero resistance through the "gap" created in the previous cycle. However, the Vipere toe model has a bilinear behav-

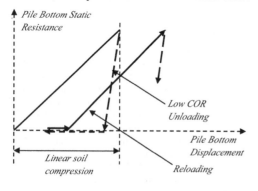

Figure 7: Vanden Berghe toe resistance model after Viking 2002

ior with a practically infinite stiffness during unloading (coefficient of restitution, COR, near zero), thereby dissipating all energy stored in the soil spring. The wave equation model, on the other hand, consumes energy only after plastification and through the associated viscous damping model.

### 8.1 Modifications to the wave equation approach

Although the rigid pile model seems to be satisfactory as long as the piles are of moderate length and the hammers of "normal" low frequency, efforts have also been made to adapt the wave equation approach to represent the vibratory hammer, pile and soil. This is reasonable for a number of reasons. First, existing computer programs offer a detailed procedure easily used by the analyst. Second, although pile elasticity is not a crucial parameter for analyzing low frequency hammers, the wave equation approach offers a rational means of analysis over a wide range of hammer frequencies and a simple approach to representing soil resistance forces. Since increasingly heavier hammers and larger piles are used, resonance effects may be more and more frequent and would go unrecognized by the rigid pile analysis. Additionally, the wave equation analysis

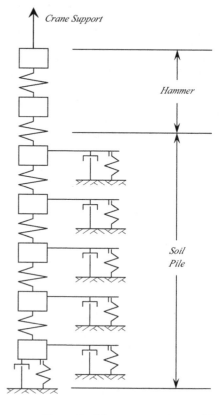

Figure 8a: The wave equation model

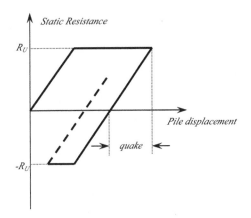

*Figure 8b: Smith static shaft resistance*

readily calculates stresses in the pile, which may be important when resonance is imminent. Most importantly, the wave equation concept has been adopted by many practioners around the world; to the practitioners it would be most convenient if the same approach could be used for vibratory analysis as for impact hammer analysis.

The wave equation model according to Smith(1960) (Figure 8a) has been described in detail in papers and manuals (e.g. GRL 1998). It's soil model includes an elasto–plastic static resistance with parameters quake and ultimate capacity. The static resistance on the shaft (Figure 8b) is assumed to be symmetric, i.e. it has during upward motions the same magnitude of resistance as during downward pile motions but with opposite sign. Thus over one cycle, the impulse of this shaft resistance is zero. The toe model is similar, yet it has no negative resistance components (Figure 8c). The Smith soil resis-

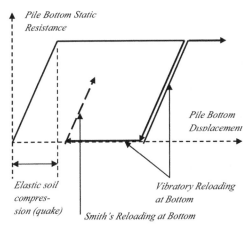

*Figure 8c: Vibratory toe model in GRLWEAP*

tance model also includes damping at shaft and toe, calculated as the product of damping parameter, static resistance and pile velocity.

Modifications necessary to produce reasonable results for vibratory pile driving simulations by GRLWEAP include the following.

1. The hammer model has to accommodate the sinusoidal forcing function over a relatively long time period. For example, a typical impact event is finished within 50 ms. In contrast, it may be necessary to analyze the vibratory motion for up to 2000 ms until a convergence in the pile variables is achieved. A long duration analysis is particularly important when analyzing a hammer with very low frequency.

2. During the analysis residual forces build up in pile and soil. These residual forces are essential for the driveability evaluation and therefore must be accurately included in the analysis. The analysis can only be stopped after the residual stresses, and therefore the pile motion, converge within a certain criterion. The residual stresses only occur where shaft resistance forces exist which exert a downward force on the pile when either the hammer applies an upward force or the pile rebounds.

3. The model must calculate power consumed by the hammer and if this value exceeds the rated value, reduction of power output must be automatically accomplished.

4. The model must include the force of the crane line and allow for extraction if this force exceeds the weight of the system. Similarly, it should be possible to analyze the pile penetration under a crowd force.

5. Instead of blow count and set per blow, the program has to calculate the rate of penetration or the time per unit penetration.

6. The end bearing has to be much more carefully modeled than for impact driving because of the separation of the pile bottom from the soil during the upward motion. Figure 8c compares the standard static toe soil resistance model according to Smith with a model that has been found to be most reasonable for vibratory analyses. In this modified static end bearing model the pile bottom will move through the gap generated in the previous cycle with zero resistance until the point of maximum displacement minus elastic rebound is reached. If the standard model were

used, the pile will work itself out of the ground unless it has a large shaft resistance. This model may be called "Residual Toe Gap". The traditional model could be referred to as a "Closing Gap". It is conceivable that the Closing Gap analysis is more reasonable than the Residual Gap model under soil conditions, such as soft clays or loose submerged sands, particularly when frequencies are low. Fortunately, the soft or loose strata do not cause high end bearing values and therefore introduce little uncertainty.

7. Energy losses are modeled with (a) Coulomb damping of the elasto-plastic static soil resistance component and (b) viscous damping. Since the particle velocities are usually lower than for impact driven piles and since the damping behavior is non-linear (Coyle et al., 1970) it is suggested to use Smith-viscous damping with damping factors, $J_{SV}$, which are double the normally suggested Smith damping factors for impact driven piles. Thus,

$$R_d = R_U \, v \, J_{SV}$$

where $R_U$ is the ultimate soil resistance at a segment, $v$ is the pile velocity, and $J_{SV}$ is the Smith-viscous damping factor.

*Table 2: Results from sensitivity study*

| Case | Q shaft | Q toe | Jshaft | J toe | Toe Cap. | Refusal Capacity |
|---|---|---|---|---|---|---|
| | mm | mm | s/m | s/m | kN | kN |
| 1 | 5 | 5 | 1.3 | 1 | 235 | 780 |
| 2 | | | 2 | 1.5 | 180 | 600 |
| 3 | | | 0.65 | 0.5 | 270 | 900 |
| 4 | 2.5 | 2.5 | 1.3 | 1 | 338 | 1125 |
| 5 | 5 | 5 | 1.3 | 1 | 90 | 860 |
| 6 | | | | | 270 | 540 |

### 8.2 *Example, driveability with Vipere and wave equation*

This example was taken from Viking (2002) and describes the driving of a sheet pile (cross sectional area 95 cm^2 and length 14m) with an ABI hammer (MRZV 800V). This unit has a variable eccentric moment with a maximum of 12 kg m and was run at 41 Hz. Below a 2.5 m thick clay layer, the soils consisted of silty sand and sand to the installation depth

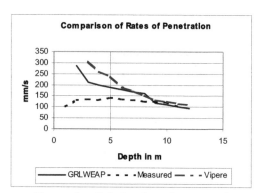

*Figure 9: Calculated and measured penetration rates after Viking (2002)*

of 12 m where the cone resistance varied between 2 and 4 MPa.

The wave equation analysis was run with typical quake values of 2.5 mm and double Smith-viscous damping values were chosen as follows: 0.33 s/m in the sand, 0.60 s/m in the silty sand and 1.2 s/m in the clay. For the toe, the damping was set to 1.0 s/m. For the sand it was assumed that the unit shaft resistance increased from 10 to 15 kPa and that the toe resistance increased from 10 to 15 kN. The driveability analysis results of the Vipere and GRLWEAP analyses are shown in Figure 9. Obviously, the calculated rates of penetration were well predicted for the 12 m depth but were grossly over predicted for the early part of the analysis by either analysis. Viking suggested that lateral motions in the early driving portion caused a reduced rate of penetration.

### 8.3 *Example, sheet pile driving*

A double sheet pile section was driven with an ICE 815 vibratory hammer as part of a cellular cofferdam installation. This hammer has an eccentric moment of 51 kg m and was run at 22.5 Hz. In order to avoid problems with misalignments, the contractor drove sheets short distances, working successively around the circular cofferdam. The soil was cohesive and relatively early refusal was the reason why measurements were taken during vibratory driving. It was concluded the soil setup along with lock friction caused the problems.

The analysis was done with a 0.7 efficiency based on field measurements of force and velocity. The calculated bearing graph is shown in Figure 10; it used double the normal quakes and double damping factors. As can be seen, stresses and capacity calculated with these parameters agreed quite well with measurements.

This example is suitable for a check on the sensitivity of some of the soil parameters of the wave equation approach. Table 2 shows capacities that were calculated when varying damping values, quakes and the percentage of the end bearing. Obviously, refusal capacities are equally sensitive to all three quantities varied with the quake reduction from 5 to 2.5 mm allowing for the greatest increase in capacity. On the other hand, with damping and quake values the same, an increase of end bearing percentage from 30 to 50% had the most pronounced effect on capacity. Clearly, for a given set of soil parameters there is a limiting end bearing (in this case around 300 kN) that cannot be overcome by the system.

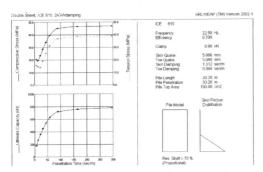

*Figure 10: Results from GRLWEAP sheet pile analysis with double quakes and double damping and comparison with measured results for (from top) compressive and tensile stress and capacity at 1600 s/m rate of penetration.*

### 8.4 *Example, reanalysis of pipe pile data*

Pile 83,1 (Table 1) was one of the better documented cases in Smart (1969) and most data needed for an analysis by GRLWEAP was available. The hammer was a Bodine resonance pile driver run at a relatively low frequency of 41 Hz and with a reduced moment of 0.94 kg m. Using standard soil resistance parameters for the silty sand and sand strata, i.e. double damping factors and 2.5 mm quakes, refusal penetrations were calculated for capacities exceeding the static weight of the driving system. On the other hand, Smart's field observations suggested a final penetration rate of more than 90 mm/s and a load test capacity of 490 kips. The static weight of the system was approximately 110 kN. The calculated displacement vs. time graph, shown in Figure 11, gives one explanation for the disagreement: the calculated dynamic double-amplitudes at the pile top and toe were only 0.57 and 0.27 mm, respectively, and therefore much less than the quake values of 2.5 mm. (Note that the constant displacement offset is

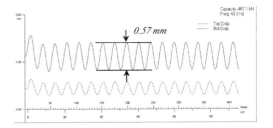

*Figure 11: Calculated displacements of Pile 83,1 Table 1a*

due to the elastic pile penetration under the static weights.) Since neither static nor dynamic forces were sufficient to fail the soil, it is not surprising that the analysis did not predict any appreciable pile penetration. This is the only example that clearly shows that the soil must have been in a reduced state of strength during driving, even though the power equations suggest that the pile bearing capacity during driving was close to that in the load test.

### 8.5 *Example, resonance calculations*

A special hammer was built and tested in the yard of the hammer manufacturer where weathered rock was encountered less than 4 m below surface. The surficial materials consisted of cohesionless soils. The hammer had an eccentric moment of 3.8 kg m and was designed for frequencies up to 80 Hz. A closed ended pipe of 170 mm diameter and 18 m length was instrumented with strain transducers and accelerometers and was rigidly bolted to the oscillator (12.4 kN weight). The bias weight was 29 kN.

One of the questions to be answered by the test was the behavior of this hammer in the neighborhood of resonance. Resonance for the pile alone on rock would occur at approximately 7 Hz (c/4L, i.e. wave speed in steel divided by 4 times the pile length); according to Poulos et al. (1980), for the above hammer weights, the resonance frequency of the hammer-pile system would be near 14 Hz.

Resonance was checked by dividing the maximum measured pile top force by the centrifugal force. The available measured data (Figure 12a) indeed suggests that the force ratio increases near the 14 and 70 Hz hammer frequencies. Unfortunately, because of a limited power rating of hammer and power pack, unlimited resonance forces could not be measured.

A check was made on the pile behavior by GRLWEAP calculating pile forces, power dissipation in the pile and force ratio. These results are shown in Figures 12b and c and very clearly indicate the same tendency as the measurements. However, in order to avoid that the program automatically reduced the power output which would have made a

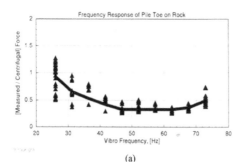

(a)

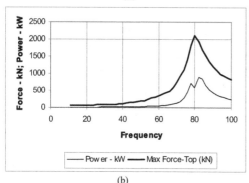

(b)

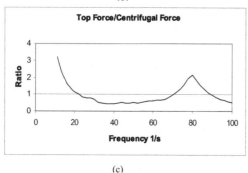

(c)

*Figure 12: a) measured force ratio b) GRLWEAP calculated forces and power transferred to top of pile; c) ratio of pile top force to centrifugal force;*

resonance check difficult, the power rating was arbitrarily set to an unrealistic high value of 1000 kW. For that reason, calculated power transfer and forces at the pile top reached much higher values than during measurements. Still, the tendency is obvious: resonance is indicated in the 15 and 80 Hz range. In fact, during the test when the hammer frequency was increased to values above 70 Hz, the stresses in the pile became so high that the pile-oscillator connection ruptured. It should be pointed out that such relatively realistic resonance studies can only be successful with an elastic pile model.

### 8.6 *Example, large caisson*

This example has been discussed by White (2002). It deals with a 12 m diameter by 0.25 m thickness concrete caisson of 25 m length. The caisson was tapered down to 0.21 m thickness at mid-length. Four APE 4B hammers with 683 kg m eccentric moment and 750 kW power each were employed with frequencies between 19.4 and 20.8 Hz. Soils consisted of clay, silty sand, sand and again silty sand with N-values of at most 3. For the analysis it was assumed that the unit shaft resistance and unit end bearing would be 10 kPa (degraded to 80% during driving) and 90 kPa, respectively. The total shaft resistance was figured for the inside and outside area of the cylinder. Damping was set to 1.3 and 1.0 s/m at shaft and toe and quakes were set to 2.5 mm. The observed final penetration times at 12 m depth ranged between 37 and 53 mm/s. The wave equation calculated penetration times are plotted vs. depth in Figure 13. They were 36 and 50 mm/s for 100% and 80% of the assumed static resistance values at the final penetration. Most of the penetration occurred prior to vibratory driving due to the static weight of hammers, transfer beams and caisson.

One of the advantages of analyzing an elastic pile model is the possibility of calculating reasonably accurate pile stresses. In the case of the concrete caisson, calculated stresses were at most 1.5 MPa. This stress is equivalent to a force of 13.8 MN. One quarter of this stress had to be transferred at each of the four points where the transfer beams were attached to the pile top.

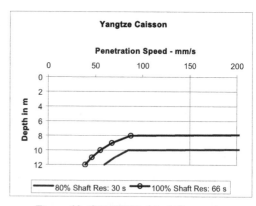

*Figure 13: GRLWEAP drivability results for a large caisson*

## 9 FINITE ELEMENT METHODS

For the sake of completeness it should be added that Leonards et al.(1995) reports, how FLAC (Fast Lagrangian Analysis of Continua), a finite element code, was adapted to the analysis of vibratory pile driving. The researchers only modeled cohesionless soils in their study. For loose sands they considered a strain hardening model and for dense soils a strain softening model. Based on the calculated stress-strain history, both shear stress and void ratio would approach the steady state values. The transverse displacements, i.e. radiation damping, in the soil were tracked as a function of time. The computer program typically required several days for an analysis.

This initial study did not get into details about vibratory pile driving specific model details. It appears that it was limited in scope due to lack of funds. Preliminary results appeared to be reasonable, and it appears that this is an interesting research tool for future studies.

## 10  SUMMARY

The following conclusions can be drawn from the examples analyzed.

- Reliability of prediction of vibratory pile driving is still elusive. There are still many unanswered question.

- The absolute magnitude of shaft resistance (or the shaft quake) is less important than the absolute magnitude of end bearing. If the percentage of shaft resistance is low then the pile can only be driven if the sum of crowd force and weight of all components exceeds the end bearing.

- The stiffness (quake) of the shaft resistance is as important as the absolute magnitude of the shaft resistance. It is possible that this parameter varies as much due to vibration as the soil resistance itself.

- The end bearing elastic properties have to be as carefully considered as the magnitude of the end bearing.

- Both static elastic and dynamic resistance parameters can have a decisive effect on driveability.

- In general, reasonable agreement between field observations and analysis can be achieved even without a liquifaction model for the shaft resistance.

- Stress predictions, even near resonance, can be made with reasonable accuracy.

## 11 RECOMMENDATIONS FOR ADDITIONAL WORK

The following studies and improvements should be made to aid the practitioner in the proper selection of vibratory piling equipment and for a better assessment of vibratory driven pile capacity.

- Most studies on vibratory pile driving deal with cohesionless soils. However, cohesive soil types pose more important questions regarding driveability than sands.

- Correlation studies dealing with the prediction of the bearing capacity of vibratory driven piles have to consider carefully the effects that the final cycles before hammer shut-off. For example, it may be necessary to require a certain procedure at the end of installation (e.g. certain final frequency while the eccentric moment is reduced to zero) to assure consistent capacity results.

- The differences between high frequency-low amplitude vibratory pile driving and lower frequency–high amplitude are still not understood and should be investigated.

- A method for calculating the increase or decrease of end bearing due to vibratory pile driving should be established for both cohesive and non-cohesive soils.

- Since liquifaction at the shaft does not seem to have a pronounced effect on drivability it is suggested to spend less effort on sophisticated resistance degradation models for the shaft and much greater efforts on investigating the change of pile toe resistance.

## 12 REFERENCES

Coyle, H.M., and Gibson, G.C., 1970. Empirical damping constants for sands and clays. ASCE, Journal of the Soil Mechanics and Foundations Division.

GRL, Goble Rausche Likins and Associates, Inc. 1998. GRLWEAP, wave equation analysis program, 4535 Renaissance Parkway, Cleveland, OH, USA.

Holeyman, A.E. and Legrand, C., 1994. Soil modeling for pile vibratory pile driving, Int. Conf. On Design and Construction of Deep Foundations, Vol. 2, pp 116-1178, Orlando, Florida, USA.

Holeyman, A.E., Legrand, C., and Van Rompaey, 1996. A method to predict the driveability of vibratory driven piles, Fifth Int. Conf. On the Use of Stress Wave Measurements on Piles, Orlando, Florida, USA.

Leonards, G.A., Deschamps, R.J., and Feng, Z., 1995. Driveability, load/settlement and bearing capacity of piles installed with vibratory hammers. Final Report submitted to the Deep Foundations Institute. School of Engineering, Purdue University, West Lafayette, Indiana, USA

Likins, G., Rausche, F., Morrison, M., and Raines, R., 1992. Evaluation of measurements for vibratory hammers. 4th Int. Conf. On the Application of Stress Wave Theory to Piles, Balkema, Rotterdam, pp 433 – 436.

Massarsch, K.R., 1992. Static and dynamic soil displacements caused by pile driving. 4th Int. Conf. On the Application of Stress Wave Theory to Piles, Balkema, Rotterdam, pp 15 – 24.

Poulos, H.G. and Davis, E.H., 1980. Pile foundation analysis and design, John Wiley and Sons, Inc.

Rausche, F., Goble, G. G. and Likins, G. 1985. Dynamic determination of pile capacity. Journal of Geotechnical Engineering, ASCE, Vol. 111, No.3, Paper No. 1951:367-383

Smart, J.D., 1969, Vibratory pile driving. Dissertation, Department of Civil Engineering, University of Illinois, USA.

Smith, E.A.L., (1960), "Pile Driving Analysis by the Wave Equation," Journal of the Soil Mechanics and Foundations Division, ASCE, Volume 86.

Viking, K., 2002. Vibro-driveability, a field study of vibratory driven sheet piles in non-cohesive soils, PhD thesis, Div. Of Soil and Rock Mechanics, Royal Inst. Of Technology, Stockholm, Sweden.

White, J., 2002. Super large caissons, lessons learned. Int. Conf. On Vibratory Pile Driving and Deep Soil Compaction, Louvain-la-Neuve, Belgium, Sept. 9-11, Balkema publisher.

# Effects of Vibratory Compaction

K. Rainer Massarsch
*Geo Engineering AB, Stockholm, Sweden*

ABSTRACT: Aspects governing the execution of vibratory compaction projects are discussed. The importance of careful planning and implementation is emphasised. Recent developments of vibratory compaction methods are presented. Design charts help to assess the suitability of soils for vibratory compaction. A hypothesis is advanced considering various factors governing the emission of vibrations from a vertically vibrating probe. It is shown that as a result of vibratory compaction, horizontal stress increase significantly and changes the stress conditions, which result in a permanent overconsolidation effect. Overconsolidation due to vibratory soil compaction is at present not taken into account in geotechnical design. The findings are illustrated with results from field measurements.

## 1 SOIL COMPACTION METHODS

Soil compaction requires geotechnical competence and careful planning on the part of the design engineer. Also the contractor must have experience from the use of vibratory compaction equipment. Each compaction method has its advantages and limitations, and thus optimal application conditions. The selection of the most suitable method depends on a variety of factors, such as: soil conditions, required degree of compaction, type of structure to be supported, maximum depth of compaction, as well as site-specific considerations such as sensitivity of adjacent structures or installations, available time for completion of the project, competence of the contractor, access to equipment and material etc.

It is increasingly common to award soil compaction projects to the lowest bidder. However, after completion of a project, this may not always turn out to be the best choice, as a too low price increases the risk that the required compaction effect is not achieved, or that the time schedule is exceeded. The compaction effect depends on several factors, which can be difficult to verify after compaction. It is thus important to apply high standards of field monitoring, quality control and site supervision during all phases of the project.

Soil compaction is a repetitive process and much can be gained from properly planned and executed compaction trials. The most important factors, which should be established and verified at the start of the project, are:

- required compaction energy at each compaction point,
- spacing between compaction points,
- duration of compaction in each point,
- ground settlements due to compaction (in compaction point and overall settlements),
- time interval between compaction passes (time for reconsolidation of soil),
- verification of the achieved compaction effect by field measurements and penetration tests,
- potential increase of compaction effect with time after compaction,
- ground vibrations in the vicinity (effects on adjacent structures and installations),
- effect on stability of nearby slopes or excavations,
- monitoring of equipment performance and review of safety aspects.

## 2 VIBRATORY COMPACTION METHODS

A variety of soil compaction methods have been developed and these are described in detail in the geotechnical literature, e.g. Mitchell (1982), Massarsch (1991), Massarsch (1999), and Schlosser (1999). In this paper, emphasis is placed on new compaction concepts. In particular, the effect of vibratory compaction on the compacted soil and of vibration propagation from the compaction probe to the surrounding soil is discussed.

Vibratory compaction methods can be classified according to the location of energy transfer from the source to the soil.

### 2.1 *Vibratory Equipment*

The first vibrators, which were developed for pile driving applications, came into use some 60 years ago in Russia. During the past decade, powerful and sophisticated vibrators have been developed for specific foundation applications, such as pile and sheet pile driving and soil compaction. These vibrators are usually hydraulically driven. Modern vibrators can generate centrifugal forces of up to 4 000 kN. The maximum displacement amplitude can exceed 30 mm. These enhancements in vibrator performance have opened new applications to the vibratory driving technique, Massarsch (1999). Recently, vibrators with variable frequency and variable static moment (displacement amplitude) have been introduced. These vibrators can be controlled electronically to adapt the vibration frequency and vibration amplitude to the varying compaction requirements.

### 2.2 *Surface Vibratory Compaction*

The compaction energy can be applied to the soil at the ground surface by steady state vibrations. The highest compaction effect is achieved in a zone close to the ground surface, but decreases usually with depth. The effective depth of compaction is difficult to assess and is influenced by a variety of factors, such as the geotechnical conditions, the type and quality of equipment, compaction procedure etc. In general, it can be assumed that the depth of influence corresponds to the diagonal length of the compaction plate. The densification effect decreases approximately linearly with depth below the centre of the compaction plate. The degree of compaction is affected by the dynamic and static force, the number of compaction cycles and the vibration frequency.

Surface compaction can be carried out with a heavy steel plate, activated by one or several powerful vibrators, cf. Fig 1.

Figure 1. Heavy vibratory compaction plate

This compaction method is being used increasingly, especially for marine and off-shore applications and has become economical due to the availability of powerful hydraulic vibrators.

Extensive investigations have been performed in connection with off-shore soil compaction projects, Nelissen (1983). It was found that the compaction effect depended on the vibration frequency and the dynamic interaction of the plate-soil system. Based on field trials on land and on the seabed, the optimal compaction parameters could be established.

Surface compaction is often used in combination with deep vibratory compaction, in order to increase the densification effect in a zone from the ground surface to approximately 3 m depth.

### 2.3 *Deep Vibratory Compaction*

The most efficient way to densify deep deposits of granular material is to introduce the compaction energy at depth, i.e in the soil layer that requires densification. The energy can either be applied by vertical or horizontal vibration, or a combination thereof. Several deep compaction methods have evolved during the past decades and are used for a variety of applications.

#### 2.3.1 *Vibro-Rod*

The Vibro-Rod method exists in several different variations, (Massarsch, 1991 and Schlosser, 1999). A compaction probe is inserted in the ground with the aid of a heavy, vertically oscillating vibrator, attached to the upper end of the compaction rod. The insertion and extraction process is repeated several times, thereby gradually improving the soil. Different types of compaction probes have been developed, ranging from conventional tubes or sheet pile profiles to more sophisticated, purpose-built tools. The Vibro-Rod method was initially developed in Japan, where a slender rod was provided with short ribs. The rod was vibrated, using conventional (often electric) vibratory pile driving equipment.

The so-called VibroWing method was developed in Sweden and is a further improvement of the Vibro-Rod method. An up to 15 m long steel rod is provided with about 0.8 to 1,0 m long radial wings, at a vertical spacing of approximately 0.5 m. The vibratory hammer is usually operated from a piling rig, Fig. 2. The frequency of the vibrator can be varied to fit the conditions at a particular site. The duration of vibration and rate of withdrawal of the probe is chosen, depending on the permeability of the soil, the depth of the soil deposit and the spacing between compaction points. The duration of compaction, the grid spacing and number of probe insertions are chosen empirically or are determined by field tests. The maximum depth of compaction depends on the capacity of the vibrator and size of the piling rig and is on the order of 10 to 15 m.

Figure 2. VibroWing method

a) MRC compaction equip- b) MRC compaction probe
ment

Figure 3. Resonance compaction method using variable frequency vibrator and flexible compaction probe

### 2.3.2 Resonance Compaction

The resonance compaction method (MRC) is similar to the Vibro-Rod method but uses the vibration amplification effect, which occurs when the vibrator, the compaction probe and the soil are vibrating at resonance. In this state, ground vibrations are strongly amplified and the efficiency of vibratory soil densification increases, cf. Fig. 3. A heavy vibrator with variable frequency is attached to the upper end of a flexible compaction probe. The probe is inserted into the ground at a high frequency in order to reduce the soil resistance along the shaft and the toe. When the probe reaches the required depth, the frequency is adjusted to the resonance frequency of the vibrator-soil system, thereby amplifying ground vibrations.

The probe is oscillated in the vertical direction and the vibration energy is transmitted to the surrounding soil along the entire probe surface. At resonance, the soil layer vibrates "in phase" with the compaction probe. At this state, vibration energy is transferred very efficiently from the vibrator to the compaction probe and to the surrounding soil, as the relative movement between the compaction probe and the soil is very small. This aspect is an important advantage, compared to conventional vibratory compaction methods. The resonance frequency depends on the dynamic and static mass of the vibrator, the mass and dynamic properties of the compaction probe and on the soil conditions. At resonance, which occurs typically between 10 and 20 Hz, the required compaction energy decreases. In this phase of soil compaction, the oil pressure of the vibrator decreases, which reduces fuel consumption and wear on the vibratory equipment.

The compaction probe is an essential component of the MRC system and is designed to achieve optimal transfer of compaction energy from the vibrator to the soil, c.f. Fig. 3a.

The probe profile has a double Y-shape, which increases the compaction influence area. Reducing the stiffness of the probe further increases the transfer of energy to the surrounding soil. This is achieved by openings in the probe, cf. Fig. 3 b. The openings also have the advantage of making the probe lighter and thereby providing larger displacement amplitude during vibration, compared to a massive probe of the same size.

Figure 4 shows how the vertical vibration velocity, measured on the ground surface, varies as a function of the vibration frequency. During probe penetration and extraction, a high vibration frequency (around 30 Hz) is used, which does not cause significant ground vibrations. During compaction, the speed of the vibrator is reduced to the resonance frequency of the probe-soil system. At resonance, ground vibrations are strongly amplified, by a factor of 3 to 4. The probe and the surrounding soil vibrate in phase, resulting in an efficient transfer of compaction energy.

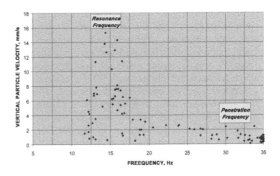

Figure 4. Vertical ground vibration velocity at a distance of 4 m from the compaction probe during probe penetration and resonance compaction

The dynamic response of the soil deposit during compaction can also be used to monitor the compaction effect. With increasing densification of the soil layers, the resonance compaction frequency rises. Also the ground vibration velocity increases and soil damping is reduced. With the aid of vibration sensors placed on the ground surface, the change in wave propagation velocity can be determined, which reflects the change of soil stiffness and soil strength, Massarsch (1995).

### 2.3.3 *Vibroflotation*

This method was invented in Germany more than 60 years ago, and its development has continued mainly there and in North America, where it was introduced in the 1940's. The equipment consists of three main parts: the vibrator, extension tubes and a supporting crane, Fig. 5.

Figure 5. Vibroflotation equipment with water jetting

Vibroflotation is the most widely used deep compaction method and extensive experience has been accumulated over the past 30 years. The vibrator is incorporated in the lower end of a steel probe. The vibrator rotates around the vertical axis to generate horizontal vibration amplitude. Vibrator diameters are in the range of 350 to 450 mm and the length is about 3 - 5 m, including a special flexible coupling, which connects the vibrator with the extension tube.

Units developing centrifugal forces up to 160 kN and variable vibration amplitudes up to 25 mm are available. Most usual Vibroflotation probes are operating at frequencies between 30 and 50 Hz. The extension tubes have a slightly smaller diameter than the vibrator and a length dependent on the depth of required penetration.

The Vibroflotation is slowly lowered to the bottom of the soil layer and then gradually withdrawn in 0.5-1.0 m stages. The length of time spent at each compaction level depends on the soil type and the required degree of compaction. Generally, the finer the soil, the longer the time required achieving the same degree of compaction. In order to facilitate penetration and withdrawal of the equipment, water jetting is utilized with a water pressure of up to 0.8 MPa and flow rates of up to 3000 l/min. The water jetting transports the fine soil particles to the ground surface and by replacing the fines with coarse material, well-compacted soil columns are obtained.

There is a fundamental difference between the Vibro-rod and the vibroflotation system. In case of the Vibro-rod (and the resonance compaction system), compaction is caused by vertically polarised shear waves, which propagate as a cylindrical wave front from along the entire shaft of the compaction probe. In addition, also horizontal compression waves are emitted, as will be discussed later. In the case of resonance compaction, a significant amount of energy can be generated at the lower end of the compaction probe (Krogh and Lindgren, 1997).

In the case of vibroflotation, the soil is densified as a result of horizontal impact of the compaction probe at the lower end. The compaction action is primarily in the lateral direction and gives rise to compression waves. Thus, it is not possible to create soil resonance using a Vibroflotation probe. The compaction zone is limited to the length of the compaction probe and the soil is improved in steps during extraction of the probe.

## 3 COMPACTABILITY OF SOILS

An important question to be answered by the geotechnical engineer at every soil compaction project is, to which degree a soil can be improved by vibratory compaction and the required compaction. Mitchell (1982) classified soils with respect to the grain size distribution. Most granular soils with a content of fines (particles < 0,064 mm) less than 10 % can be compacted by vibratory and impact methods. The disadvantage with an assessment of compaction based on the grain size distribution is that a large number of soil samples are required to obtain a realistic picture of the geotechnical conditions. Due to the loose state of the soil prior to compaction, it is difficult and costly to retrieve representative soil samples. The compaction is also affected by soil layering, which may not be apparent from the inspection of a limited number of soil samples. It is therefore preferable to assess the compatibility by the cone penetration test, CPT.

With CPT, detailed and reliable information of the soil strength and the soil layering is obtained. Massarsch (1991) has proposed that the compactability of soils can be based on the cone resistance and on the friction ratio, Fig. 6.

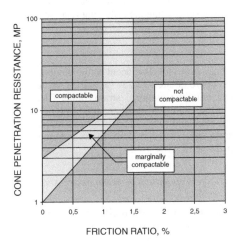

FRICTION RATIO, %

Fig. 6. Soil classification for assessment of deep compaction based on CPT (Massarsch, 1991).

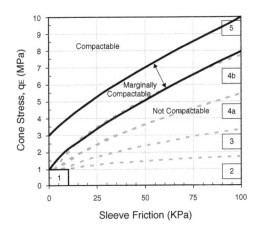

Figure 7. Soil classification for deep compaction with soil type boundaries per Eslami-Fellenius (1997; 2000)

The diagram assumes uniform soil conditions. Layers of silt and clay can reduce, however, the effectiveness of the compaction. The CPTU, where the excess porewater pressure is measured, can also be used to determine the soil stratification and the occurrence of less permeable silt and clay layers.

Fig. 7 presents the classification boundaries proposed by Massarsch (1991) together with soil type boundaries proposed by (Eslami and Fellenius, 1997; 2000) in a soil classification chart with the cone stress as a function of the sleeve friction.

The numbered areas in Fig.7 denote soil type, as follows:

1 = Very Soft Clays or Sensitive Soils
2 = Clay and/or Silt
3 = Clayey Silt and/or Silty Clay
4b = Sandy Silt and Silt
4a = Fine Sand and/or Silty Sand
5 = Sand to Sandy Gravel

There is good agreement with between the two charts, Fig. 6 and 7. However, Fig. 7 is preferred as it covers a wider range of soils and also indicates soil type.

# 4 COMPACTION MECHANISM IN SAND

Although extensive information is available in the literature concerning the application of different soil compaction methods, little is mentioned about the mechanism, which causes the rearrangement of soil particles and densification.

An attempt is made to describe the factors, which are considered important for the densification process. To obtain a better understanding of the compaction process, it is necessary to consider the stress strain behaviour of granular soils.

## 4.1 *Energy transfer from compaction probe to soil*

Different types of energy sources can be used for soil densification. However, the basic mechanism, governing the energy transfer from the vibrating source to the surrounding soil, is in principle similar. The resonance compaction probe is used as an example, as the vibration energy is mainly transmitted to the surrounding soil along the shaft of the probe. However, a similar approach can also be used for assessing other compaction methods.

An important question for the prediction of ground vibrations caused by vibratory compaction is, whether there exists an upper limit to the vibration energy, which can be transmitted from the probe to the surrounding soil. In the plastic zone at the interface between the soil and the compaction probe, the maximum shear stress can approximated by

$$\tau_f = v_{max} \, Z_s = v_{max} \, C_s^* \, \rho \qquad (1)$$

where $v_{max}$ is the maximum particle vibration velocity, $Z_s$ is the soil impedance and $\rho$ is the bulk density. The soil impedance is the product of the strain-dependent shear wave velocity, $C_s^*$ and the soil density $\rho$. Similar relationships can be used to assess the energy transfer at the base of the probe, Bodare and Orrje (1988). According to Eq. 1, the maximum vibration velocity, which can be transmitted to the soil in the plastic zone, can be estimated from

37

$$v_{max} = \frac{\tau_f}{C_s^* \, \rho} \qquad (2)$$

It should be noted that the shear wave velocity $C_s^*$ at large strains is significantly lower than the small-strain shear wave velocity, which is determined by seismic field tests. In Fig. 8, the reduction of the shear modulus with shear strain is shown Massarsch (1983). The tests were performed in a resonant column apparatus on a sample of dry sand of medium density. The shear modulus at different strain levels is divided by its maximum value to obtain a modulus reduction factor. From the shear modulus and the soil density, the shear wave velocity can be readily determined. The shear wave reduction factor is the square root of the shear modulus reduction factor. At a strain level of about 1 %, where the sand can be assumed to behave plastically, the shear wave velocity is only about 25 % of the maximum value. In the case of a medium dense sand, the shear wave velocity decreases thus from around 150 m/s to about 40 m/s. Thus, also the soil impedance decreases in the plastic zone – a fact that is generally neglected. This strain effect must be taken into consideration when assessing energy transfer from a vibration source to the surrounding soil.

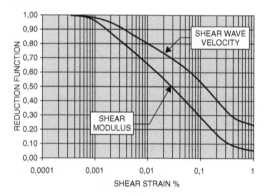

Figure 8. Reduction of shear modulus and shear wave velocity as a function of shear strain in a saturated sand

Assuming medium dense sand, a shear wave velocity at large strain of $C_s^*$ of 40 m/s and a bulk density of 2 t/m³, it is possible to estimate from Eq. 2 the maximum vibration velocity which can be transmitted to the surrounding soil. The shear strength can be estimated from e.g. sleeve friction measurements using the CPT, and is in medium dense sand typically on the order of 100 kPa. The calculated maximum particle vibration velocity, which can be transmitted to the soil, is thus 1.2 m/s. This maximum vibration velocity can now be used to estimate vibration attenuation from the compaction probe.

An interative process can be used to estimate the shear strain level g, using the following relationship

$$\gamma = \frac{v_{max}}{C_s^*} \qquad (3)$$

The shear strain level adjacent to the compaction probe is approximately 1,25 % and is in good agreement with the assumed strain value. Clearly, the soil in the vicinity of the compaction probe is in the plastic state and it would be erroneous to use the small-strain shear wave velocity for calculating the soil impedance.

### 4.2 Vibration propagation from the source to the surrounding soil

Adjacent to a vertically oscillating compaction probe, three compaction zones can be identified:

1. elastic zone: where the shear strain level is below 10-3 %, and no permanent deformations can be expected,

2. elasto-plastic zone: where the strain level ranges between 10-3 and 10-1 %, where some permanent deformations will occur, and

3. plastic zone: where the soil is in a failure condition and is subjected to large strain levels >10-1 %.

These three zones are indicated schematically in Fig. 9. Also shown is the assumed attenuation of the vibration velocity (particle velocity) of the cylindrical wave front in the ground. In the plastic zone the vibration velocity is relatively constant and limited by the shear strength of the soil. The vibration amplitude attenuates rapidly in the elasto-plastic zone. In the plastic, and the elasto-plastic zone, the wave propagation velocity is strain-dependent and increases with distance from the energy source. In the elastic zone, the wave propagation velocity is constant, due to the limitation by the shear strength of the soil.

### 4.3 Horizontal ground vibrations

In the geotechnical literature is often assumed that in the case of vertically oscillating probes or piles, only vertical ground vibrations occur. However, in addition to a vertically polarised shear wave, which is emitted along the shaft of the compaction probe, also horizontal vibrations are generated. These are caused by the friction between the compaction probe and the soil, and cause horizontal stress pulses. These are directed away from the probe during its downward movement. The horizontal stress changes result in a compression wave and increased lateral earth pressure. This aspect will be discussed in the following sections.

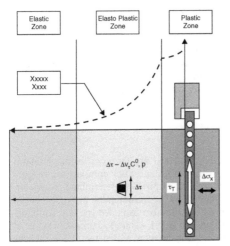

Figure 9. Transfer of vibration energy from the compaction probe to the surrounding soil

Fig. 10 shows the results of field measurements during vibratory compaction using the MRC system, Krogh & Lindgren (1997). Horizontally oriented vibration sensors (geophones) were installed at different levels below the ground surface, 2,9 m from the centre of the compaction probe. At the time of the vibration measurements, the tip of the compaction probe had passed the lowest measuring point.

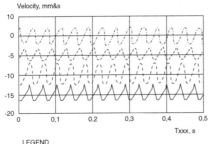

LEGEND
-·-·-· Channel 1: Horizonal surface geophone (mm's)
- - - - Channel 2: Horizontal underground geophone (mm's) Depth: 1.65 m
- - - Channel 3: Horizontal underground geophone (mm's) Depth: 3.55 m
——— Channel 4: Horizontal underground geophone (mm's) Depth: 5.95 m

Figure 10. Horizontal vibration amplitude measured during resonance compaction, from Krogh & Lindgren (1997)

In spite of the vertically oscillating compaction probe, strong horizontal vibrations are generated. These were of the same order of magnitude as the vertical vibration amplitudes. It will be shown that as a result of vibratory compaction, the horizontal stresses increase in the soil. This compaction effect is of great importance as it changes permanently the stress conditions after compaction.

## 5 INCREASE OF LATERAL STRESS

An aspect of vibratory compaction, which is not generally appreciated, is the increase of the lateral stresses in the soil due to vibratory compaction. Sand fills (such as hydraulic fill) are usually normally consolidated prior to compaction. The lateral earth pressure increases significantly as a result of vibratory compaction, as shown by the measured sleeve resistance, Massarsch and Fellenius (2002). The sleeve friction $f_s$ can be approximated from Equation 4

$$f_s = K_0 \, \sigma'_v \tan\left(\phi'_a\right) \tag{4}$$

where $\sigma'_v$ = effective vertical stress, $K_0$ = earth pressure coefficient, $\phi'_a$ = the effective sleeve friction angle at the soil/CPT sleeve interface. The ratio between the sleeve friction after and before compaction, $f_{s1}/f_{s0}$ can be calculated from Eq. 5

$$\frac{f_{s1}}{f_{s0}} = \frac{K_{01} \, \sigma'_{v1} \, \tan\left(\phi'_{a1}\right)}{K_{00} \, \sigma'_{v0} \, \tan\left(\phi'_{a0}\right)} \tag{5}$$

where $f_{s0}$ = sleeve friction before compaction, $f_{s1}$ = sleeve friction after compaction, $K_{00}$ = coefficient of earth pressure before compaction (effective stress), $K_{10}$ = coefficient of lateral earth pressure after compaction (effective stress), $\sigma'_{v0}$ = vertical effective stress before compaction, $\sigma'_{v1}$ = vertical effective stress after compaction, $f_{a0}$ = sleeve friction angle before compaction, $f_{a1}$ = sleeve friction angle after compaction.

If it is assumed that the effective vertical stress, $\sigma'_v$, is unchanged by the compaction, the ratio of the lateral earth pressure after and before compaction, $K_{01}/K_{00}$ can then be estimated from the relationship according to Eq. 6

$$\frac{K_{01}}{K_{00}} = \frac{f_{s1}}{f_{s0}} \frac{\tan\left(\phi'_{a0}\right)}{\tan\left(\phi'_{a1}\right)} \tag{6}$$

Equation 6 shows that the earth pressure coefficient is directly affected by the change of the sleeve friction and of the friction angle of the soil. The horizontal stresses can vary significantly within the compacted soil. The highest horizontal stresses are expected close to the compaction points and decrease with increasing distance. The initial stress anisotropy initiates a stress redistribution, which can to some extent explain the change of soil strength and of the stiffness with time.

### 5.1 Case History

Extensive field investigations were carried out in connection with a major land reclamation project,

Fellenius and Massarsch (2001). CPTU tests were performed in a loose deposit of hydraulic fill, as well as at different time intervals following resonance compaction. Figure 11 shows the results of cone resistance and sleeve friction measurements prior to, and after compaction. The average of several CPT measurements has been used to determine the increase of cone resistance and sleeve friction.

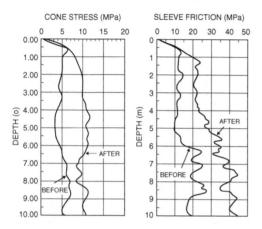

Fig. 11. Filtered average values of cone stress and sleeve friction from before and after compaction, Fellenius and Massarsch (2001).

The cone stress and the sleeve friction increased in the sand deposit as a result of the vibratory compaction. On average, the cone stress is doubled or higher, indicating efficient densification of the sand fill. The specifications requirement of a cone stress of at least 10 MPa was satisfied. The effect of vibratory soil compaction on the stress conditions is also evidenced by increase in sleeve friction, on average about 2.5 times, which is about the same increase ratio as that of the cone stress. Thus the friction ration after compaction remained almost unchanged. This observation is in good agreement with experience reported in the literature and suggests, that in these cases, the horizontal stress has been increased.

The friction angle after compaction was not determined, but it is assumed that it is about 36°, which results in a sleeve friction ratio of 0.8. Inserting this ratio and the ratio of sleeve friction of 2.5 into Eq. 6 gives a ratio of earth pressure coefficient of 2.0. Because the earth pressure coefficient prior to compaction, $K_{00}$, can be assumed to be 0.5, the earth pressure coefficient after compaction, $K_{01}$, is 1.0.

## 6 OVERCONSOLIDATION EFFECT

For many geotechnical problems, knowledge of the overconsolidation ratio is important. Empirical relationships have been proposed for the coefficient of lateral earth pressure of normally and overconsolidated sands and for the overconsolidation ratio, OCR,

$$\frac{K_{01}}{K_{00}} = OCR^m \tag{7}$$

where $K_{00}$ and $K_{01}$ are the coefficient of lateral earth pressure before and after compaction, respectively and m is an empirically determined parameter. Schmertmann (1985) recommended m = 0.42, based on compression chamber tests. Mayne and Kulhawy (1982) suggested m = $1 - \sin(\phi)$. Jamiolkowski et al (1988) found that the relative density, $D_R$, influences m and that m varied between 0.38 and 0.44 for medium dense sand ($D_R = 0.5$). Figure 12 illustrates the relationship from Eq. 7, which shows that even a modest increase of the lateral earth pressure increases the overconsolidation ratio significantly.

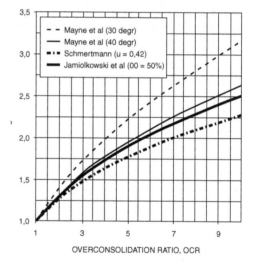

Fig. 12. Relationship between overconsolidation ratio and ratio of earth pressure coefficients for overconsolidated and normally consolidated sand, Fellenius and Massarsch (2001).

Sleeve resistance measurements reported in the literature and the above shown field tests show that the ratio $f_{s12} / f_{s01}$ varies between 1,5 and 3,5, Massarsch and Fellenius (2002). If it is assumed that the effective friction angle increases due to compaction from on average 30 to 36 degrees, $K_{01}/K_{00}$ ranges according to Eq. 6 between 1.2 – 1.8. An average value of $K_{01}/K_{00} = 1.6$ yields an overconsolidation ratio OCR according to Eq. 7 and Fig. 12 in the range of 2.5 – 4.0. This overconsolidation effect, which is generally neglected, is important for the analysis of many geotechnical problems.

## 6.1 *Change of Stress Conditions*

The stress conditions in loose, water-saturated sand will undergo a complex change of stress conditions during vibratory compaction. Energy is transmitted from the compaction probe to the surrounding soil at the tip as well as along the sides of the probe. The transmitted vibration energy depends on the capacity of the vibrator, the shear resistance along the probe and on the shape and size of the probe.

At the beginning of compaction of loose, water-saturated sand, the stress conditions will correspond to that of a normally consolidated soil. When the soil is subjected to repeated, high-amplitude vibrations, the pore water pressure will gradually build up and the effective stress is reduced. During the initial phase of compaction, the soil in the vicinity of the compaction probe is likely to liquefy. Whether or not liquefaction will occur, depends on the intensity and duration of vibrations and the rate of dissipation of the excess pore water pressure. If the soil deposit contains less permeable layers (e. g. silt and clay), these will increase the liquefaction potential. At liquefaction, the effective stresses and thus the shear strength of granular soils are zero. Although the probe continues to vibrate, the soil will not respond as only little vibration energy can be transmitted from to the soil. With time, the excess pore water pressure will start to dissipate. The rate of reconsolidation will depend on the permeability of the soil (and interspersed layers).

Figure 13 illustrates the change of effective stresses in a dry granular soil, which is subjected to repeated compaction cycles. During vibratory compaction, high oscillating centrifugal forces (loading and unloading) are generated (up to 2,000 kN) that temporarily increase and decrease the vertical and the horizontal effective stress along the compaction probe and at its tip. The initial stresses of the normally consolidated soil correspond to point (A). During the first loading cycle, the stress path follows the $K_{00}$-line to stress level (B). Unloading to stress level (C) occurs at zero lateral strain and horizontal stresses remain locked in. Each reloading cycle increases the lateral earth pressure, which can reach the passive earth pressure. At the end of compaction, stress point (D) is reached. The vertical overburden pressure is the same after compaction but the horizontal effective stresses have been increased. The lateral earth pressure after compaction can reach the passive value, $K_p$. The dynamic compaction has thus caused preconsolidation and increased the horizontal effective stress. The increase of the sleeve friction and the high lateral earth pressure as measured in the above presented case history, Fig. 11 can thus be explained by Fig. 13.

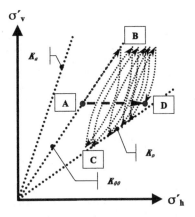

Fig. 13. Stress path of soil in the vicinity of a compaction probe during tow compaction phases; before (A), during first compaction phase (B, C, D and E) and during second compaction phase (E, F, G).

In the opinion of the author, Fig. 12 illustrates important aspects of vibratory compaction. The change of the stress conditions from a normally consolidated state to an overconsolidated state is influenced by several factors, such as the compaction method, the state of stress state prior to compaction and the strength and deformation properties of the soil. At MRC compaction, the vertically oscillating probe generates (as a result of friction between the probe and the soil) high, horizontally oscillating force, which is responsible for the high lateral earth pressure in the soil after compaction.

## 7 ACKNOWLEDGEMENTS

The stimulating discussions and astute comments by Dr. Bengt H. Fellenius during the development of the vibratory compaction concept are gratefully acknowledged. The support by Krupp GfT Tiefbautechnik in preparing this paper is acknowledged.

The research into developing the vibratory compaction concept was funded by the Lisshed foundation, which is acknowledged with gratitude.

## 8 SUMMARY

New developments in vibratory compaction have been made possible as a result of more powerful and sophisticated equipment. In spite of this positive development, many vibratory compaction projects are designed and executed without sufficient knowledge and understanding of the principles, which govern deep soil compaction.

The process of soil compaction using the Vibro-rod system is fundamentally different to that of Vibroflotation. In case of the Vibro-rod, compaction is due to shear waves and compression waves, which are transmitted from the shaft of the compaction probe to the surrounding soil. The Vibroflotation method uses compression waves to compact the soil. The resonance compaction method uses the vibration amplification effect to increase compaction efficiency.

Reliable charts are available to assess the compactability of soils. These are based on results of cone penetration tests, CPT with sleeve friction measurements.

A hypothesis is proposed which explains the mechanism of energy transfer from the compaction probe to the surrounding soil. An upper limit exists of the vibration amplitude, which can be transmitted in the plastic zone adjacent to the compaction probe.

As a result of vibratory compaction, high lateral stresses are created, which cause a permanent over-consolidation effect. This aspect should be taken into consideration when calculating settlements in vibratory-compacted soils.

## REFERENCES

Bodare, A., and Orrje, O., 1985. Impulse load on a circular surface in an infinite elastic body -Closed solution according to the theory of elasticity. Rapport 19, Royal Institute of Technology (KTH), Stockholm, Sweden, 88 p.

Eslami, A., and Fellenius, B.H., 1997. Pile capacity by direct CPT and CPTu methods applied to 102 case histories. Canadian Geotechnical Journal, Vol. 34, No. 6, pp. 880  898.

Fellenius, B.H., and Eslami, A., 2000. Soil profile interpreted from CPTU data. Proceedings of the International Conference "Year 2000 Geotechnics", Bangkok, November 27 30, 2000.

Fellenius, B. H. & Massarsch. K. R., 2001. Deep compaction of coarse-grained soils - A case history. 2001 - A Geotechnical Odyssey: The 54th Annual Canadian Geotechnical Conference. Paper submitted for publication, 8 p.

Jaky, J. 1948. Earth pressure in silos. Proceedings 2$^{nd}$ International Conference on Soil Mechanics and Foundation Engineering, ICSMFE, Rotterdam, Vol. 1, pp. 103  107.

Jamiolkowski, M., Ghionna, V. N, Lancelotta R. and Pasqualini, E., (1988). New correlations of penetration tests for design practice. Proceedings Penetration Testing, ISOPT-1, DeRuiter (ed.), Balkema, Rotterdam, ISBN 90 6191 801 4, pp 263  296.

Krogh, P. and Lindgren, A., 1997. Dynamic field measurements during deep compaction at Changi Airport, Singapore, Examensarbete 97/9. Royal Institute of Technology (KTH), Stockholm, Sweden, 88 p.

Massarsch, K.R., 1991. Deep Soil Compaction Using Vibratory Probes. American Society for testing and Material, ASTM, Symposium on Design, Construction, and Testing of Deep Foundation Improvement: Stone Columns and Related Techniques, Robert C. Bachus, Ed. ASTM Special Technical Publication, STP 1089, Philadelphia, pp. 297  319.

Massarsch, K.R., 1994. Settlement Analysis of Compacted Fill. Proceedings, 13th International Conference on Soil Mechanics and Foundation Engineering, ICSMFE, New Delhi, Vol. 1, pp. 325 - 328.

Massarsch, K.R., 1999. Deep compaction of granular soils. State-of-the art report, Lecture Series: A Look Back for Future Geotechnics. Oxford & IBH Publishing Co. Pvt. Ltd. New Delhi & Calcutta, pp. 181 – 223.

Massarsch, K. R. & Broms, B. B., 2001. New Aspects of Deep Vibratory Compaction. Proceedings, Material Science for the 21$^{st}$ Century, JSMS 50$^{th}$ Anniversary Invited Papers, Vol. A, The Society of Material Science, Japan, pp. 172 – 179.

Massarsch, K. R. & Fellenius, B. H., 2001. Vibratory Compaction of Coarse-Grained Soils. Canadian Geotechnical Journal, Vol. 39. No. 3, 25 p.

Mayne, P.W. and Kulhawy, F.H. (1982). $K_0$-OCR relationship in soil. ASCE Journal of Geotechnical Engineering, 108 (6), pp. 851- 870.

Mitchell, J.K., 1982. Soil improvement-State-of-the-Art, Proceedings, 10th International Conference on Soil Mechanics and Foundation Engineering, ICSMFE, Stockholm, June, Vol. 4., pp. 509  565.

Nelissen, H. A. M., (1983). Underwater compaction of sand-gravel layers by vibration plates. Proceedings ECSMFE Helsinki, Volume 2, pp. 861 – 863.

Schlosser, F., 1999. Amelioration et reinforcement des sols (Improvement and reinforcement of soils). 4$^{th}$ International Conference on Soil Mechanics and Foundation Engineering, ICSMFE, Hamburg, 1997, Vol. 4, pp. 2445-2466.

Schmertmann, J.H., 1985. Measure and use of the in situ lateral stress. Practice of Foundation Engineering, A Volume Honoring Jorj O. Osterberg. Edited by R.J. Krizek, C.H. Dowding, and F. Somogyi. Department of Civil Engineering, The Technological Institute, Northwestern University, Evanston, pp. 189-213.

*Modeling of vibratory driving*

*Vibratory Pile Driving and Deep Soil Compaction - TRANSVIB2002,*
*Holeyman, VandenBerghe & Charue (eds.), © 2002 Swets & Zeitlinger, Lisse, ISBN 90 5809 521 5*

# A Mechanical Model for the investigation on the Vibro-drivability of Piles in Cohesionless Soils

Robert O. Cudmani, Gerhard Huber & Gerd Gudehus
*Institute of Soil and Rock Mechanics, University of Karlsruhe*

A mechanical model based on the wave propagation method was developed to investigate the drivability of piles in granular soils. The soil resistance is divided into shaft and toe Resistance and its evolution is modelled by mathematical expressions, which are based on experimental results and justified from a soil mechanics point of view. The parameters of the model are estimated from impact penetration or vibro-penetration tests. The correct description of the toe resistance, which is essentially different from the toe resistance for static and impact penetration, was found to be the key for the modelling of vibrodriving. Considering the evolution of the vertical toe force and displacements during loop, two different penetration modes, called "cavitation" and "no-cavitation", have been identified, depending on the set of machine parameters (frequency, static load and static moment) and on the initial state of the soil. It is shown that model is able to predict the occurrence of both penetration modes and the transition between them for different machine parameters.

## 1 INTRODUCTION

Vibratory driving is a dynamic method to install piles, tubes or sheet piles into the ground. Instead of impact, the technique uses a harmonic, vertical excitation force generated by a vibratory driver fixed to the head of the pile. The vibratory driver basically consists of a vibrator, a bias mass, isolation springs between the bias mass and the vibrator and a connection (usually hydraulic clamps) of the vibrator to the pile. The vibrator produces vertical forces by means of counter-rotating eccentric masses at frequencies usually ranging from 5 Hz to 100 Hz. Typical displacement amplitudes and dynamic force amplitudes of commercial vibrators range from 5 to 25 mm and from 200 to 2500 kN, respectively.

While the basic principles of vibratory driving are relatively simple and well known, there are a number of essential still open questions. This holds particularly for the parameters affecting the driving resistance, the soil disturbance that is generated in the vicinity of the pile, and the bearing capacity of a vibro-driven pile. As experiments show, the driving resistance depends not only on the soil properties and the soil state, but also on the parameters of the vibro-driver-pile system (VERSPOHL (14), DIERSSEN (6), SCHMID (12), RODGER & LITTLEJOHN (11)). The physical quantities controlling the force displacement amplitude of the vibrator frequency $f$, the vibrator mass $M$, the bias weight $F_{sta}$ and the static moment of the eccentric masses with respect to its rotation axes $S_v$ as well as those quantities defining the dynamic characteristics of the pile, i.e. its mass, geometry and mechanical properties.

Different approaches have been proposed to predict the drivability of piles taking into account the influence of the above-mentioned quantities. In the *parametric* approaches mostly based on field experience, the vibro-drivability of a pile is judged by empirical rules. The *energy balance* methods are based on the energy conservation principle, i.e. the energy transferred into the vibro-driver-pile-soil system is equal the sum of the energy consumed to overcome soil resistance and the energy loss caused by dissipation. In order to determine the energy to drive the pile, it is assumed that the soil behaves in an ideally plastic manner. An empirical factor is used to determine the dissipated energy as a fraction of the input energy. Therefore, the penetration rate is a function of the bearing capacity, the dissipation factor and the energy released by the vibrator.

In the *rigid body* methods, the vibro-driver-pile-soil system is modelled by a one-dimensional mechanical model with one degree of (Figure 1). The simplification is only valid if the driving frequency is much lower than the vertical resonance frequencies of the pile, since only then the pile behaves as a rigid body. The forces acting on the pile are the harmonic, and the static driving force (action), and the soil resistance (reaction). The latter results from non-linear springs and viscous dashpots at toe and the shaft. Using Newton's Second law ($\Sigma F = ma$) the penetration rate is obtained by numerical integration of the acceleration $a$ over time. In the *wave-equation* methods, the pile is assumed to behave like an elastic rod. The penetration rate is obtained by numerical integration of the one-dimensional wave propagation

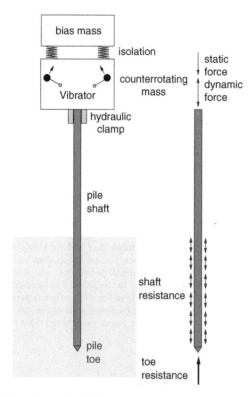

bias mass

isolation

Vibrator

counterrotating mass

static force
dynamic force

hydraulic clamp

pile shaft

shaft resistance

pile toe

toe resistance

Figure 1. Mechanical modeling of vibratory driving.

equation. Similar to the rigid body method the soil resistance is also modelled by springs and dashpots.

The last two approaches allow us to consider the influence of machine parameters and pile properties in drivability predictions rather precisely. However, since the influence of the soil is taken into account in a rather simple way, the quality of the predictions strongly depends on the ability of the toe and shaft resistance models to predict the actual soil response.

Most of the available models are based on the assumption that the soil resistances during impact and vibratory driving are similar. However, this assumption is invalid, since the penetration mode during vibratory driving strongly depends on the machine parameters and the soil state, and can be rather different from the one during impact driving (RODGER & LITTLEJOHN (11), DIERSSEN (6), CUDMANI (4)). Even if the soil resistance is reasonably modelled, the application of a particular model for drivability predictions depends on an accurate procedure for the estimation of the model parameters.

The aim of this work is to improve and verify the mechanical model proposed by DIERSSEN (6) for the simulation of vibratory driving of piles in cohesionless soils, and to provide a procedure to evaluate the model parameters. The experimental investigations of vibratory driving, on which our mechanical model is based, will be briefly presented in Section 2. The soil resistance

at the toe will be considered in Section 3. After modelling the shaft resistance in Section 4 and the damping due to wave radiation in Section 5, a method for the determination of the model parameters will be proposed in Section 6. Finally, an application of the model will be shown in Section 7.

## 2 EXPERIMENTAL INVESTIGATION OF VIBRATORY DRIVING

Our experimental investigation of vibratory driving included both laboratory tests with a model pile (D=0.036 m, L=3.0 m) (CUDMANI (4) and full-scale field tests with a steel pile (D=0.160 m, L=7.25 m) (CUDMANI & HUBER (5)). The field tests were performed in four different sites in South-West Germany (Karlsruhe, Berghausen, Hochstetten). The ground of the test sites consists of cohesionless soil layers ranging from silt to fine gravel, lying in a loose to dense state. In the field tests the machine parameter were systematically varied to investigate their influence on the soil resistance. The laboratory tests were carried out a calibration camber under controlled boundary and initial conditions, using a vibro-penetration testing (VPT) equipment (CUDMANI & HUBER (5)) and a quartz sand as soil material. The aim of the laboratory tests was to study the influence of the initial soil state on the vibratory driving for given machine parameters. Some of the test results will be shown and discussed in the next sections.

## 3 MODELLING OF TOE RESISTANCE

The toe resistance during vibratory driving is characterized by two different modes of penetration: *cavitation* and *no-cavitation* (CUDMANI & HUBER (5)). The cavitation mode takes place if the contact between the toe and the underlying soil is lost during the upward phase of the pile motion. Whereas if the contact between the soil and the pile is not lost in a driving loop, the no-cavitation mode occurs. One the driving loop of the cavitation and no-cavitation modes can be divided into four and two phases. Adopting the nomenclature from CUDMANI & HUBER (5) the expressions for the description of the toe resistance during cavitation and no-cavitation driving are:

*cavitation mode*

Phase I (1-2) $\qquad F_s = F_s(u_1) - K_e(u - u_1)$

switch condition: $F_s = 0$, $\dot{u} < 0$

Phase II (2-3) $\qquad F_s = 0$

switch condition: $\dot{u} = 0$, previously $\dot{u} < 0$

Phase III (3-4) $\qquad F_s = 0$

switch condition: $\dot{u} < 0$ and $u = u_4$

Phase IV (4-q') $\qquad F_s = K_b (u - u_4)$

switch condition: $\dot{u} = 0$, previously $\dot{u} > 0$

46

*no-cavitation mode*

Phase I (1-2)          $F_s = F_s(u_1) - K_e(u - u_1)$

switch condition: $\dot{u} = 0$, previously $\dot{u} < 0$

Phase II (2-1)         $F_s = K_e(u - u_2)$

switch condition: $F = K_{s,max}$

Phase II (1-1′)        $F_s = K_{s,max}$

switch condition: $\dot{u} = 0$, previously $\dot{u} > 0$

In addition, the full-scale tests show that for particular settings of the machine parameters, other modes of penetration, which combine the features of the cavitation and no-cavitation modes, can also occur. Figure 2 shows two of such transition modes. As we can see, they occur when the contact between the toe and the underlying soil gets lost during the upward motion of the pile, but is immediately restored after the pile motion reverses. Based on the experimental results of our full-scale tests, the following assumptions are made to incorporate the transition modes in the mathematical description of the toe resistance.

1. The maximum toe resistance $F_{s,max}$ in the no-cavitation mode cannot be exceeded
2. Transition modes occur when the motion reversal takes place between the points 2 and 4 (see Figure 5a in CUDMANI & HUBER (5) AND Figure 3).

3. The nearer to point 2 the reversal occurs, the more the stiffness during downward motion resembles the one during the upward motion, and the larger is the ratio of the plastic deformation to the total downward displacement of the soil (see equation 3 in CUDMANI & HUBER (5)).

Fig. 3 shows qualitatively a possible evolution of the toe resistance predicted by the model when the bias weight is increased while keeping the other machine parameters constant. At the beginning the toe resistance develops according to the cavitation mode (Fig. 3a). The maximum toe force is lower than $F_{s,max}$. After increasing

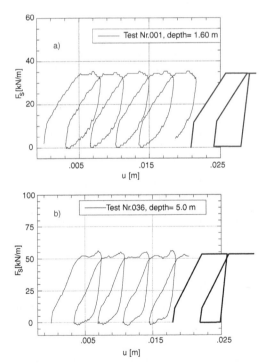

Figure 2. Transition modes: a) test site Hagenbach, b) test site Hochstetten.

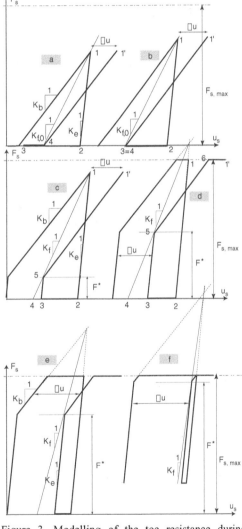

Figure 3. Modelling of the toe resistance during vibratory driving: a), b) cavitation mode; from c) to e) transition modes; f) no-cavitation mode.

the bias weight the fastest penetration velocity for the cavitation mode is achieved since the motion reversal occurs at point 4 (Fig. 3b). A further increase of the bias weight causes reversal between points 2 and 4, i.e the toe resistance shows a transition between cavitation and no-cavitation (Fig. 3c). After the reversal, the soil response is stiffer until the line with the inclination $K_f$ is touched. If we continue increasing the bias weight, the maximum toe resistance $F_{s,max}$ is achieved (Fig. 3d, e). At the end, if the bias weight is heavy enough the penetration resistance develops according to the no-cavitation mode (Fig. 3f). Similar changes of the toe resistance can occur if the frequency, the static moment or the vibrating mass are changed.

For the evaluation of the inclination $K_f$ which is necessary for determining the position of point 4, the following interpolation function is proposed:

$$K_f = K_{f,0} + (K_e - K_{f,0}) \frac{F_s^*}{F_{s,max}} \qquad (1)$$

For the cavitation mode $F_s^* = 0$ (see Figure 3) and the inclination $K_{f,0}$ is obtained from $K_b$ and $\beta_p$ (see equation 3 in CUDMANI & HUBER (5)) as:

$$K_{f,0} + \frac{K_b}{1 - \beta_p} \qquad (2)$$

## 4 SHAFT RESISTANCE

For the description of the shaft resistance the model proposed by DIERSSEN (6) is used. Assuming that the soil behaviour is rate independent and that no adherence exists between the soil and the pile, the shear stress $\tau$ at the shaft can be modelled by the following function:

$$\tau = (\tau_0 - \tau_{max} \, \text{sgn} \, (\dot{u})) \exp \left[ - \frac{C_1}{\tau_{max}} (u - u_0) \right.$$

$$\left. \text{sgn} \, (\dot{u}) \right] + \tau_{max} \, \text{sgn} \, (\dot{u}) \qquad (3)$$

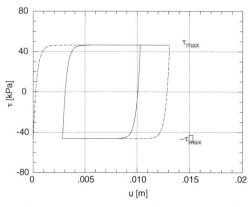

Figure 4. Shear stress at the shaft as simulated with equation 3.

$\tau_{max}$ is the maximum shear stress at a particular depth, $\dot{u}$ is the current velocity and $C_1$ is a constant. The maximum shear stress is defined according to the Coulomb friction law by $\tau_{max} = p_N \tan\delta$ ($p_N$: pressure on the shaft, $\delta$: interface friction angle). The quantities $\tau_0$ and $u_0$ are the shear stress and the shear displacement achieved at the previous motion reversal. Figure 4 shows the evolution of the shear stresses according to equation 3 (compare with the experimental curve, Figure 9 in CUDMANI & HUBER (5)).

## 5. DAMPING DUE TO WAVE RADIATION

Due to the radiation of waves from the toe and the shaft, a part of the energy released by the vibrator is dissipated. The exact evaluation of the radiated energy would require the solution of a complex non-linear wave propagation problem. For this reason, and since radiation damping commonly does not play an important role in drivability analysis (DIERSSEN (6)), we simplify the problem and use a suitable analytical solution. The energy radiation at the toe and at the shaft are treated separately. In addition, it is assumed that the radiated waves propagate as in a linear elastic half space (shear stiffness $G$, density $\rho$ and Poisson coefficient $\nu$). For modelling of the radiation damping at the toe the equivalent viscous damping force (according to LYSMER & RICHARD (8)) for a circular footing (radius $r$) resting on an elastic half space is applied.

$$D_s = \frac{3.4}{1 - \nu} r^2 \, \dot{u} \sqrt{G \, \rho} \qquad (4)$$

For modelling of the radiation damping at the shaft the equivalent damping shear stress proposed by MEYNARD & CORTÉ (9) is applied:

$$\tau_d = \dot{u} \sqrt{G \, \rho} \, \min \left( 1, (1 - \frac{\tau}{\tau_{max}} \, \text{sgn} \, (\dot{u})) \right). \qquad (5)$$

The original expression of MEYNARD & CORTÉ (9) is only valid for sticking motion and does not include the empirical *sliding* factor $\min(1, (1 - \tau/\tau_{max}\text{sgn}(\dot{u})))$ in equation 5. This was introduced by DIERSSEN (6) to take into account the fact that the wave radiation at the shaft diminishes if sliding takes place instead of sticking. The factor is equal to one after a reversal ($\tau = -\tau_{max} \, \text{sgn}(\dot{u})$) when sticking dominates, and vanishes at the limit state ($\tau = \tau_{max} \, \text{sgn}(\dot{u})$) in which the shaft slides passed the soil at constant shear stress.

## 6 DETERMINATION OF MODEL PARAMETERS

The applications of the mechanical model presented in the previous sections requires the determination of four parameters for the modelling of the toe resistance ($K_b$, $K_e$, $K_{f,0}$ and $F_{s,max}$), three parameters for the description of the shear stresses at the shaft ($p_N$, $\delta$ and $C_1$) and the elasticity parameters ($G$, $\nu$) for damping due to wave radiation.

## 6.1 Toe resistance

The parameters $K_b$, $K_e$, $K_{f,0}$ and $F_{s,max}$ depend on the diameter of the pile. For the determination of the parameters from empirical relationships or field testing methods, it is useful to deal with quantities which do not depend on the pile geometry. Assuming that the displacement fields induced in the underlying soil by two piles of diameters $d_1$ and $d_2$ are geometrically similar and that the similarity factor is equal to the ratio of the diameters, it is easy to derive the diameter independent parameters $C_b$, $C_f$ and $C_e$ and $q_s$, viz.

$$C_i + K_{i,1}\frac{d_1}{A_1} = K_{i,2}\frac{d_2}{A_2}, \text{ with } i = b, f, e \qquad (6)$$

$$q_{s,max} = \frac{F_1}{A_1} = \frac{F_2}{A_2} \qquad (7)$$

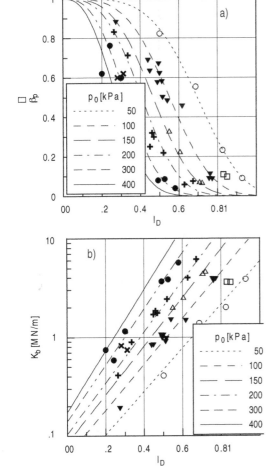

Figure 5. Dependence of a) the plastic deformation ratio $\beta_p$ and b) the inclination $K_b$ on the initial state of the soil for the VPT-penetrometer.

Knowing $\beta_p$ and $C_b$ we can get the inclination $C_{f,0}$, which is the inclination $C_f$ for the cavitation mode from expression 2. The plastic deformation ratio $\beta_p$ depends on the soil type and state. Figure 5 shows the relationship between $\beta_p$ and the state variables determined for medium quartz sand in the calibration chamber using the VPT-device (CUDMANI (4)). It is seen that the values of $\beta_p$ can vary considerably depending on the relative density and the penetration depth represented by $p_0$. The values of $\beta_p$ for the test site Hochstetten (sand and fine gravel, relative density $I_D \approx 0.5$ to 0.6) varied from ca. 0.35 to 0.45 (DIERSSEN (6)). For the test site Hagenbach (sandy silt, $I_D \approx 0.3$ to 0.4, the values of $\beta_p$ varied from ca. 0.35 to 0.6 (CUDMANI (1)). The field values of $\beta_p$ are slightly lower than those shown in Figure 5a.

Similar to $\beta_p$, the inclination $C_b$ also depends on the soil type and the initial state. Figure 5b shows the dependence of $K_b$ on the state variables determined for the VPT-penetrometer in a calibration chamber ($C_b = K_b$ $D_{VPT}/A_{VPT} = K_b$ 0.036/0.001). An empirical correlation between $C_b$ and the blow count $N_{10}$ of dynamic probing tests (DPT) could be determined on the basis of the results of the full-scale field tests. We obtained the relationship $C_b = a (N_{10})^2$, where $a \approx 1000$ kN/m² for the heavy (DPH) and $a \approx 400$ kN/m² for the light penetrometer (DPL). Different from $\beta_p$, the values of $C_b$ in the field and laboratory tests show a good correlation.

For the evaluation of $C_e$ the empirical relationship $C_e \approx 3$ to $5$ $C_{f,0}$ can be used. For $C_e > 5$ $C_b$ the influence of $C_e$ on the evolution of the toe resistance is negligible.

As a first approximation, $q_{s,max}$ can be set equal to the static penetration resistance $q_c$. This value can be directly obtained from CPT or estimated from correlations between $q_c$ and the results of other field testing methods. Alternatively, $q_c$ can be estimated from the initial soil state using empirical or analytical correlations. For instance, CUDMANI & OSINOV (3) proposed a method based on the spherical cavity expansion problem (OSINOV & CUDMANI (10)) to evaluate $q_c$ in cohesionless soils.

The procedure for the estimation of the model parameters outlined above presupposes that they do not depend on the machine parameters. Unfortunately, the parameters can actually show some dependence on the machine parameters. Many of the uncertainties associated with the empirical approach can be eliminated if a vibro-penetration testing technique (VPT) as proposed by CUDMANI & HUBER (5) is used for parameter determination. The advantage of this technique is that the parameters can be determined under dynamic loading conditions which are very similar to those during pile driving.

## 6.2 Shaft Resistance

The angle of friction $\delta$ increases with the ratio of the shaft roughness to the mean grain size. $\delta$ can be taken as a fraction of the critical angle of friction of the soil or determined by shear box tests. The pressure $p_N$ is obtained form the expression:

$$p_N = K\sigma'_z. \qquad (8)$$

Here, $\sigma'_z$ is the initial vertical effective stress and $K$ is an earth pressure coefficient. The value of $K$ depends on two different mechanisms. On one hand, the displacement on the soil during the passage of the pile toe causes and increase in the horizontal stress, i.e an increase in $K$ similar to the one during the expansion of a cylindrical or a spherical cavity OSINOV & CUDMANI (10)). On the other hand, the alternating shearing of the soil in the vicinity of the shaft causes densification with a reduction of the horizontal stresses (CUDMANI & GUDEHUS (2), HAUSER (7), VIKING (15)) or loosening with an increase in the horizontal stresses (TEJCHMAN (13), WEHR (16)) depending on the tendency of the soil skeleton to contract or to dilate during cyclic shearing. The experimental values of $K$ for the full-scale tests vary approximately from $K_a$ to $2K_0$ for loose and dense soil states ($K_a$ = active earth pressure coefficient, $K_0$=coefficient of earth pressure at rest).

With a known displacement $u$ needed to mobilize a particular shear stress $\tau$, the constant $C_1$ is evaluated from equation 3. For instance, if after a displacement of 1% of the pile diameter ($d_p$) 95% of $\tau_{max}$ should be mobilized, then:

$$C_1 = \frac{\ln(1 - 0.95)}{0.01 \, d_p} \tau_{max} \qquad (9)$$

## 7 IMPLEMENTATION OF THE MODEL

The proposed soil resistance model can be used in conjunction with both the rigid-body and the wave-propagation methods. Since the wave-propagation method is more general, we chose this approach for implementation and verification of the model. The problem of the one-dimensional wave propagation in an elastic rod is described by the second order differential equation

$$\rho_p \, A_p \, \ddot{u} + E_p \, A_p \, \frac{\delta^2 u}{\delta^2 x} + (\tau + \tau_d) \, U_p = 0 \qquad (10)$$

with the boundary conditions

$$E_p \, A_p \left( \frac{\delta u}{\delta x} \right)_{x=0} = F_{sta} + S_v \, (2\pi f)^2 \, \text{co}(2\pi f t) \quad (11)$$

$$E_p \, A_p \left( \frac{\delta u}{\delta x} \right)_{x=1} = F_s \, (t) \quad (12)$$

The quantities $U_p$, $E_p$ and $\rho_p$ are the perimeter, the elastic modulus and the density of the pile. For the integration of the differential equation 10, an implicit finite element scheme was used. The finite element discretization and the boundary conditions are schematically shown in Figure 6.

## 8 AN APPLICATION EXAMPLE

In order to show the application of the model, some typical experimental results of full-scale tests in Hangenbach will be calculated in this section.

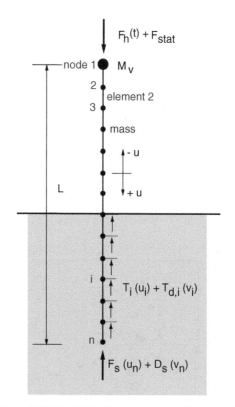

Figure 6. FE-Model for the numerical simulation of pile driving.

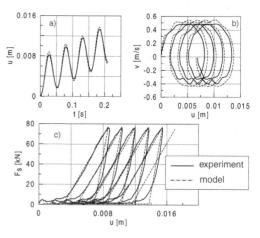

Figure 7. Comparison of experimental and numerical results for Test 001 at 6.0 m depth.

The ground of the test site consists of loose to medium dense sandy and silty layers. The depth of ground water table is 2 m. In the vibratory driving tests, a Müller MS15H vibratory driver was used with a maximum centrifugal force of 500 kN at a maximum frequency of 50 Hz. The static moment could be varied from 3 to 15 kgm (free-vibration amplitude from 2 mm to 10 mm). The vibratory driver had a total mass of 2000 kg (1500 kg vibrating and 500 kg non vibrating). A bias load was generated by a Müller MS-M 10000T telescope leader with a maximum force of 50 kN.

The machine parameter used in the three considered tests were:

| test | $f$ [Hz] | $S_v$ [kgm] | $F_v$ [kN] | $F_{sta}$ [kN] |
|------|------|------|------|------|
| 001 | 22 | 5 | 90 | 17 |
| 050 | 24 | 3 | 65 | 45 |
| 056 | 35 | 3 | 145 | 34 |

The model parameter were determined according to the procedure proposed in Section 6. The parameters of the toe resistance are:

| test | $z$ [m] | $C_b$ [MPa] | $C_f$ [MPa] | $C_e$ [MPa] | $F_{s,max}$ [kN] |
|------|------|------|------|------|------|
| 001 | 6.0 | 174 | 290 | 695 | 100 |
| 050 | 2.0 | 84 | 139 | 333 | 90 |
| 056 | 3.6 | 63 | 104 | 250 | 55 |

For the determination of $\tau_{max}$, the values $K = 1$, $\delta = 30°$ and $\gamma = 16$ kN/m$^3$ above and $\gamma' = 8$ kN/m$^3$ beneath the ground water table were adopted.

The comparison of numerical and experimental results at the toe are presented in Figures 7, 8, 9. The diagrams (a) show the evolution of the pile displacement with time, and diagrams (b) are the corresponding phase diagrams. The evolution of the toe resistance with the toe displacement is shown in the diagrams (c). Figure 10 shows measured and predicted penetration rates in different depths for test 001. The coincidence between the experimental and the numerical results is

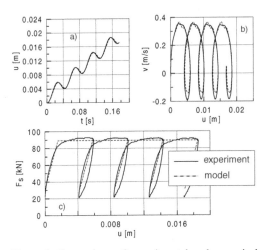

Figure 8. Comparison of experimental and numerical results for Test 050 at 2.0 m depth.

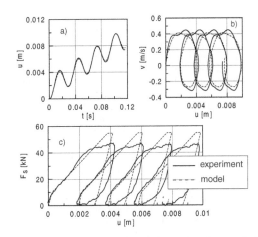

Figure 9. Comparison of experimental and numerical results for Test 056 at 3.6 m depth.

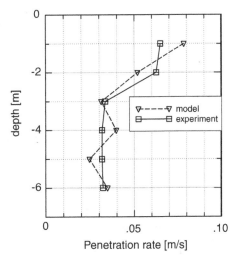

Figure 10. Comparison of experimental and numerical penetration rates for test 001 at depths 1.0, 2.0, 3.0, 4.0, 5.0, and 6.0 m.

very good. In particular the model is able to predict the occurrence of cavitation, no-cavitation and transition modes.

9 CONCLUSIONS

A mechanical model for the description of vibro-driving of piles in granular soils was presented. The penetration resistance was separated into toe and shaft resistances. The energy dissipation due to wave radiation was modelled through velocity dependent forces at the toe and at the shaft. The correct modelling of the toe resistances requires the use of a relatively complicated model. Penetration modes with and without cavitation as well as the transition between them are modelled. The required constants depend on the properties and the state of the soil and can be easily estimated with a procedure presented above or by the vibro-penetration testing technique. The shaft resistance is modelled using an exponential function, the calibration of which requires the estimation of the horizontal earth pressure acting on the pile shaft during driving and the determination of the angle of friction of the interface. The suitability of the model for the investigation of the vibro-drivability of piles was shown in an application example. The application of the model can be extended to piles having sections other than circular by giving the effective toe and shaft areas on which the soil resistance will act. An extension to sheet piles is possible by adding the additional penetration resistance originated in the clutch between the sheet piles.

REFERENCES

1. R. Cudmani: A Soil Mechanical Model for Modelling Vibratory Driving in Cohesionless Soils. *Workshop Vibratory Driving*, Karlsruhe, 1997 (in German)
2. R. Cudmani, G. Gudehus: Settlements of Sand due to Cyclic Twisting of a Tube. *Proceedings* of the IUTAM Symposium on Theoretical and Numerical Methods in Continuum Mechanics of Porous Materials, Stuttgart, pp. 387–396, 1999
3. R. Cudmani, V. Osinov: The Cavity Expansion Problem for the Interpretation of Cone Penetration and Pressuremeter Tests. *Canadian Geotechnical Journal*, **38**, pp. 622–638, 2001.
4. R. Cudmani: Static, Alternating and Dynamic Penetration of Cohesionless Soils. *Veröffentlichung des Institutes für Bodenmechanik und Felsmechanik der Universität Karlsruhe*, Nr. 152, 2001 (in German)
5. R. Cudmani, G. Huber: Development of a Vibro-Penetration Test (VPT) for the In-situ Investigation of Cohesionless Soils. *Proceeding* of the International Conference on Vibratory Driving and Deep Compaction, 2002 (in this volume)
6. G. Dierssen: Ein Bodenmechanisches Modell zur Beschreibung des Vibrationsrammens in Körnigen Böden. *Veröffentlichungen des Instituts für Bodenmechanik und Felsmechanik der Universität Fridericiana in Karlsruhe*, Heft 133, 1994 (in German)
7. A. Hauser: Entwicklung und Anwendung eines Experimentellen Modells zur Untersuchung von Alternierender Scherbeanspruchung in Sand. *Diplomarbeit*, Institut für Bodenmechanik und Felsmechanik der Universität Fridericiana in Karlsruhe, 2001.
8. J. Lysmer, F.E. Richart: Dynamic Response of Footings to Vertical Loading. *Journal of the Soil Mechanic and Foundations Division*, ASCE, 92(1), S. 65–91, 1966.
9. A. Meynard, J.F. Corté: Experimental Study of Lateral Resistance During Driving. *Preceeding* of the 2nd International Conference on the Applications of Stress-Wave Theory on Piles, S. 210–220, 1984.
10. V. Osinov, R. Cudmani: Theoretical Investigation of the Cavity Expansion Problem Based on a Hypoplasticity Model. *Journal of numerical and analytical methods in Geomechanics*, 25, pp. 473–495, 2001
11. A.A. Rodger, G.S. Littlejohn: A study of Vibratory Driving in Granular Soils. *Géotechnique*. 30(3), pp. 269–293, 1980
12. W.E. Schmid: Driving Resistance and Bearing Capacity of Vibro-driven Piles. In: *Performance of Deep Foundations* ASTM STP 444, pp. 362–375, 1969
13. J. Tejchman: Modelling of Shear Localisation and Autogeneous Dynamic Effects in Granular Bodies. *Veröffentlichungen des Institutes für Bodenmechanik und Felsmechanik der Universität Fridericiana in Karlsruhe*, Heft 140, 1997.
14. J. Verspohl: Periodenverdoppelung bei einem Schwinger mit Hysteretischer Kennlinie. *Veröffentlichung des Institutes für theoretische Mechanik*, Universität Karlsruhe, 1990
15. K. Viking: Driveability Studies of Vibro-driven Piles in Non cohesive Soils. Licentiate Thesis 2029. Division of Soil and Rock Mechanics. Department of Civil and Environmental Engineering, Royal Institute of Technology, Stockholm, 1998
16. W.C.S Wehr: Granulatumhüllte Anker und Nägel – Sandanker. *Veröffentlichungen des Institutes für Bodenmechanik und Felsmechanik der Universität Fridericiana in Karlsruhe*, Heft 146, 1999

*Vibratory Pile Driving and Deep Soil Compaction - TRANSVIB2002,*
*Holeyman, VandenBerghe & Charue (eds.), © 2002 Swets & Zeitlinger, Lisse, ISBN 90 5809 521 5*

# Vibratory pile driving analysis. A simplified model

J.-G. Sieffert
*ENSAIS – IMFS UMR 7507 CNRS-ULP, Strasbourg, France*

ABSTRACT: For the design of vibro-driving, the more important questions are : is vibro-driving adapted to the mechanical characteristics of the soil ? Will the vibrator be able to drive the pile to the desired depth ? The answer at these both questions suppose the designer know the effects of the vibration on the shaft resistance and on the toe resistance. The paper describes a simplified model in order to understand better the behaviour of the global system vibrator-pile-soil. The software "BRAXUUS" is used to compare experimental and numerical data obtained with this model. The prediction of refusal depth so as the effects of the quake are also analyzed. It is essential to have in the future a better understanding of the effects of the amplitudes of displacement on the surrounding soil.

## 1 INTRODUCTION

For the design of vibro-driving, a very important question is the selection of the best-adapted vibrator in accordance with pile and soil characteristics. This requires the knowledge of the interaction between pile and soil. Everybody agrees the effect of the vibrations is a decrease of the static shaft resistance, and perhaps also a decrease of the static toe resistance. But the exact physical behaviour is not well known. It is however possible to have a good adjustment between measured and calculated results with a model using a lot of parameters. Our choice is to analyze a model with very few parameters.

### 1.1 *Assumptions*

We suppose the pile is rigidly fasten to the vibrating mass (including the eccentric masses) through the clamp. The additional mass (or suppressor) is supposed not to interfere the dynamic movement and has only a static action. Consequently, we have to distinguish the total vibrating mass $M_V$ and the static mass $M_{stat}$ which includes the additional mass.

Like other authors in the past, we consider the pile is a rigid body. Contrary to pile driving, effects of wave propagation are not very important in vibro-driving. We can consider the dynamic behaviour like a succession of steady states, more precisely, a succession of periodic states. The amplitude of the displacement is well known for a section of a pile supposed free at the bottom and without soil effects (free hanging system for example) is given by Equation 1:

$$U(y) = U(0)\cos\frac{\omega y}{c} \qquad (1)$$

where $y$ = the distance of the considered section from the free end, $U(0)$ = the amplitude of the displacement at the free end of the pile, $\omega$ = the circular frequency and $c$ = the velocity of the longitudinal waves in the pile given by Equation 2:

$$c = \sqrt{\frac{E}{\rho}} \qquad (2)$$

where $E$ = the elastic modulus and $\rho$ = the unit mass of the pile. Of course, $U(0)$ depends on the mechanical characteristics of the vibrator. Consequently, for a free "system" the ratio amplitude at the top - amplitude at the bottom of the pile is not subordinated to the vibrator characteristics. Table 1 shows the values of this ratio versus the length $L$ of a steel pile ($E$ = 2.1 $10^5$ MPa, $\rho$ = 7800 kg, frequency = 30 Hz). It is clear that until a length of about 15 m, the ratio can be considered close to 1: all sections of the pile have more or less the same amplitude of displacement. This validates the assumption of a rigid body.

Table 1. Ratio amplitude at the top - amplitude at the bottom of a pile

| $L$ (m) | $U(L)/U(0)$ |
|---|---|
| 10 | 0.935 |
| 15 | 0.855 |
| 20 | 0.747 |
| 25 | 0.615 |

Only two forces are opposed to the penetration during vibro-driving: the shaft and the toe resistances. A plastic behaviour is considered for both forces. Figure 1 gives explicitly the assumptions concerning these forces.

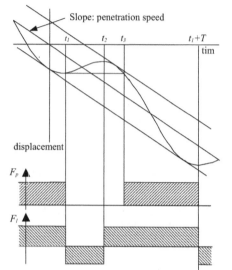

Figure 1. Shaft and toe resistances versus time and displacement

- from time $t_1$ to time $t_2$, the pile goes up: the shaft resistance $F_1$ is turned down and there is no force $F_p$ at the bottom,
- from time $t_2$ to time $t_3$, the pile goes down but its depth is smaller than the maximum depth reached during the preceding cycle: the shaft resistance is turned up and there is no force at the bottom,
- from time $t_3$ to time $t_1 + T$ ($T$ = period of the vibrator) the pile continues to go down. The maximum depth during the preceding cycle is reached: both toe and shaft resistances are turned up.

The times $t_i$ are functions of the velocity of the system which can be separated into two parts: the penetration speed $V_f$ (slope of the increase of depth during one cycle) and of the velocity of the alternative displacements.

Vibro-driving is especially adapted to sandy soils for which viscosity effects are not pronounced: we do not consider effects of the velocity on the soil behaviour law in terms of viscosity. On the other hand, our choice is to have as few parameters as possible: it is ever possible to adjust a model with a lot of parameters, but generally the meaning of each parameter can be not very clear.

### 1.2 Motion

The vibrator is reduced to a vibrating mass and an alternative force produced by the rotation of the unbalanced masses. The motion is given by Equation 3:

$$M_V \frac{d^2z}{dt^2} = P_S - Q + M_e\,\omega^2 \sin \omega t \qquad (3)$$

where $M_V$ is the vibrating mass including the clamp and the pile, $P_S$ the static weight of the "system" vibrator-clamp-pile, $Q$ the resistance of the soil, and $M_e\,\omega^2$ the amplitude $F_{exc}$ of the alternative load (see Fig. 2).

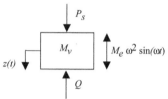

Figure 2. Forces on the vibrating mass

This non linear equation is solved using Runge-Kutta's method by means of our software BRAXUUS able to calculate at one and the same time the alternative motion during each cycle and the penetration curve depth versus time.

## 2 FIELD TESTS ANALYSIS

We had at our disposal some field tests data "SIP-DIS" performed near Hamburg. We present here only the analysis of test n°5 which was performed with a vibrator MS10 and a sheet pile PU16 into a sandy soil. Table 2 gives the values of key features of the vibrator and of the sheet pile.

Table 2. Vibrator and sheet pile features

| Vibrator MS 10 (including clamp) | |
| --- | --- |
| vibrating mass | 1960kg |
| static weight | 25.6 kN |
| frequency | 29.5 Hz |
| force | 171.8kN |
| Sheet pile PU16 | |
| length | 15 m |
| weight | 11.2 kN |
| transversal area | 95.2 cm$^2$ |
| perimeter | 1.66 m |
| vibrating mass | 1120 kg |
| total vibrating mass $M_V$ | 3 080 kg |
| total static weigh $P_S$ | 36.8 kN |

In order to justify the assumption concerning the sheet pile, the amplitudes of the displacements at the top and at the bottom of the sheet pile were calculated considering the sheet pile as elastic and rigid. The system was supposed free (free hanging). Table 3 shows that the amplitudes for the rigid body is close to the average of the amplitudes at the top and at the bottom for an elastic body (4%) for which all sections have also the same phase.

Table 3. Amplitudes of displacements

| sheet pile supposed | elastic | rigid |
|---|---|---|
| amplitude at the top (mm) | 1.562 | 1.623 |
| amplitude at the bottom (mm) | 1.818 | 1.623 |

## 2.1 Pulling test

The field test includes two parts: first a penetration test and then a extraction test. We begin with the pulling test analysis (see Figure 3): we will have to consider only the shaft resistance because the toe resistance is nil during extraction.

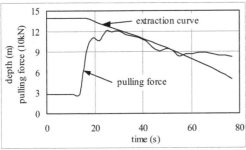

Figure 3. Extraction curve – pulling force

An external static pulling force is applied to the vibrator. Figure 3 shows clearly that this force is not constant during the test. In addition, it is not sure that this force is perfectly vertical. Consequently buckling behaviour can affect the sheet pile and disturbs the friction along the sheet pile. Nevertheless we have taken into account the variability of the pulling force in our software and calculated the shaft resistance in the form of $\alpha\, z^2$ ($z$ is the penetration depth) at the beginning of the extraction.

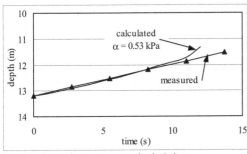

Figure 4. Comparison measures and calculation

Calculated and measured curves are very close for $\alpha$ = 0.53 kPa as shown on Figure 4, at least for a depth between 13.2 and 11.9 m.

Classical calculations of static active normal force applied by a soil on a retaining wall are based on Equation 4:

$$P = 0.5\, K_{\alpha\gamma}\, \gamma\, s\, z^2 \qquad (4)$$

where $K_{\alpha\gamma}$ is the earth pressure factor (typically about 0.3 for a sandy soil), $\gamma$ the specific weight (about 20 kN/m$^3$) and $s$ the perimeter (here 1.66 m). So a first approach gives for the normal static earth resulting force a coefficient about 4.98 kPa. Comparing with the obtained value of $\alpha$, we can accept the friction force during vibro-driving as a part of the static normal force (here about 10%).

## 2.2 Penetration test

As said before the sheet pile was driven at first in the soil with the same eccentricity moment and the same frequency. A static retaining force was applied to the vibrator. In the same way as for the extraction test, we give on Figure 5 the penetration curve and the retaining static force.

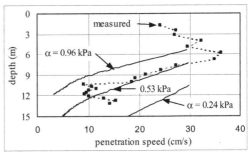

Figure 5. Penetration curve – retaining force

Here also we cannot consider that this force is constant during the test. The variability has been introduced in our software and also the shaft resistance in the same form of $\alpha z^2$. First calculations were performed without toe resistance. The Figure 6 shows the penetration speed calculated for three values of $\alpha$.

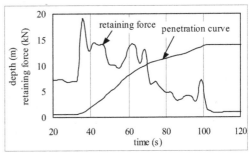

Figure 6. Penetration speed versus depth

It is clear that the theoretical results obtained for $\alpha$ = 0.53 kPa are close to the experimental data, at least for a depth between 6 and 12 m.

The next step is the introduction of a toe resistance. The selected depth at time $t = 0$ is the depth corresponding to the static equilibrium of the weight, the shaft and the toe resistance. Calculations were performed for $\alpha = 0.53$ kPa with two values of the toe resistance: 10 kN (1.05 Mpa) and 20 kN (2.10 Mpa).

On Figure 7 we see that the results are satisfying from 7m depth for a toe resistance of 10 kN and using the measured static weight. The results are furthermore better with a toe resistance of 20 kN and using a constant static weight of 26 kN.

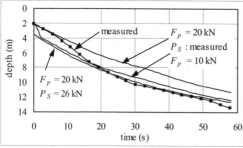

Figure 7. Penetration curves ($\alpha = 0.53$ kPa)

CPT tests were performed at the Hamburg test site. In fact the soil consists of layers: a sand layer from 0 to 6m depth (toe resistance of about 15 MPa), a silt layer from 6 to 10m depth (toe resistance of about 2 MPa), and a dense sand layer beyond 10m (toe resistance of about 20 MPa). Our calculations were performed supposing an homogeneous soil. An explanation of the difference between the calculated and the measured curves can be the non-homogeneity of the soil. On the other hand, the comparison of the toe resistance during the vibro-driving (1 to 2 MPa) and the CPT (about 20 MPa below 10m) shows that the vibrations reduce the static toe resistance in a factor of about 5 to 10%.

### 2.3 Conclusions

We can conclude that the simplified model with only two parameters and a rigid pile is able to accurately describe a field penetration test. In addition, the penetrability can be predicted using a small portion of the static resistances (about 10% for this example).

### 3 REFUSAL DEPTH

An other important concern is to be able to predict correctly the refusal depth defined as the depth for which the mean depth does not increase (penetration speed nil). Unfortunately none test at our disposal meets this definition.

### 3.1 Existing criteria

We will use dimensionless parameters:

$$X = \frac{F_l}{P_S} \qquad Y = \frac{F_p}{P_S} \qquad A = \frac{F_{exc}}{P_S} \qquad (4)$$

Vié (2001) considers the same modeling for the pile and for the soil as we do and divides the X, Y space into three zones:
- zone a: without displacements, limited by:

$$X \geq A - 1 \quad \text{and} \quad Y \geq 1 + A - X \qquad (5)$$

- zone b: alternating displacements, limited by:

$$X \leq A - 1 \qquad (6)$$

- zone c: continuously going down, limited by:

$$X \geq A - 1 \quad \text{and} \quad 1 - X \leq Y \leq 1 + A - X \qquad (7)$$

Figure 8 shows these zones for $A = 10$.

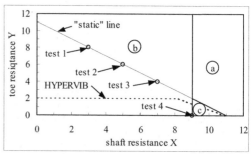

Figure 8. Diagram $X - Y$ (for $A = 10$)

From a physical point of view, the crossing from zone c to zone a is not clear, as well as the crossing from zone b to zone a through zone c (for $Y = 0$ for example).

HYPERVIB (after Vié 2001) suggests that the same diagram X-Y is restricted by:

$$\text{if } X \leq \frac{A-1}{1+\theta} \text{ then } Y = 2 \quad \text{else } X = \frac{2A-Y}{2(1+\theta)} \qquad (8)$$

This limit calculated with $\theta = 0.1$ as advised is given on Figure 8. On the same figure, we give also the "static" line corresponding to:

$$X + Y = A + 1 \qquad (9)$$

Equation 9 is obtained by writing that the sum of the amplitudes of the resistance forces equals the sum of the amplitudes of the active forces. This writing is valid for static forces and the use easy, but of course, this equation does not make sense for dynamic motions and is given here only as information.

### 3.2 Numerical results

In order to have $A = 10$, we have modified some previous parameters, and will use yet these given in Table 4. A lot of numerical tests were performed.

We present here only some results. Details on these tests are given in Table 5, and the corresponding points are showed in the diagram (Fig. 8).

Table 4. Vibrator and sheet pile features

| frequency | 25 Hz |
| eccentric moment | 15.903 m kg |
| alternating force amplitude | 392.4 kN |
| total vibrating mass | 3080 kg |
| total static weight | 39.24 kN |

Table 5. Numerical tests

| test | 1 | 2 | 3 | 4 |
| --- | --- | --- | --- | --- |
| X | 3 | 5 | 7 | 9 |
| Y | 8 | 6 | 4 | 0 |
| penetration speed (cm/s) | 6.9 | 7.0 | 3.5 | 2.4 |

In Figures 9 and 10 the displacements and the velocities of the "system" over one period of the vibrator are respectively represented. The "system" is stationary when the velocity and the acceleration are nil at the same time. For test 1, the "system" is non-stopping. For test 2, its pause is very short, and for test 3, we have 2 pauses. For test 4, the pause is large, and contrary to the other tests, the "system" goes only down and does not go up.

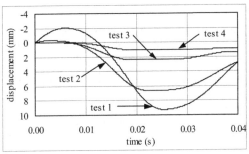

Figure 9. Displacement versus time

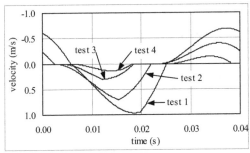

Figure 10. Velocity versus time

In comparison with the diagram $X - Y$ (Fig. 8), tests 1, 2 and 3 are in Zone b and we verify that we have effectively alternating displacements for these tests. On the other hand, the diagram $X - Y$ does not distinguish the displacements without pause from the

displacements with one or two "stop and go". Test 4 is in Zone c and we verify that the system is continuously going down, but with small displacements and velocities. An other continuously going down motion can exist when the alternating velocity is ever smaller than the penetration speed. This case in not described in the diagram $X - Y$. We see also that the limit given by HYPERVIB ($Y = 2$) is too pessimistic.

It is very easy to calculate the amplitude of the "free" system. This amplitude (peak to peak) is given by Equation 10:

$$a_0 = 2 \frac{M_e}{M_V} = 10.3 \, \text{mm} \qquad (10)$$

where $M_e$ is the eccentric moment. For our tests the amplitudes are reduced to 0.67 $a_0$ for test 2, 0.23 $a_0$ for test 3 and 0.10 $a_0$ for test 4. This analysis shows also that the penetration speed and the amplitude of displacement decrease in the same time. At the end, both are simultaneously nil. From a physical standpoint the amplitudes cannot be too small: the energy necessary to reduce the soil friction is transmitted to the soil through these amplitudes. Therefore a reasonable refusal depth criterion has to satisfy two conditions at the same time: a nil penetration speed and alternating displacement amplitudes adequately large.

Our model using rigid-plastic behaviour for both shaft and toe resistances does not satisfy these two conditions. It is necessary to add something at the initial refusal depth definition: it can be a limitation for the displacements amplitudes, for the transmitted energy or even for the penetration speed.

### 3.3 "Stop and go" criterion

An other definition of the refusal depth can be based on "stop and go" phases. Figure 11 shows the beginning of these phases in diagram $X - Y$.

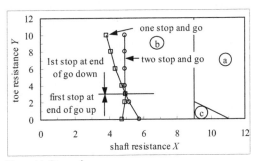

Figure 11. Stop and go

We can notice a significant change for $Y = 3$ for both functions. For the two "stop and go" phase, X decreases when $Y$ increases until $Y = 3$ and then is constant ($X = 4.92$). It is clear that the refusal depth corresponds to the zone limited by the curve two "stop and go" and the limit between zone b and zone a. This zone is large: we need other information on the behaviour of the "system".

### 3.4 Energy analysis

The transmitted energy from system to the soil can be an other way to define the refusal depth.

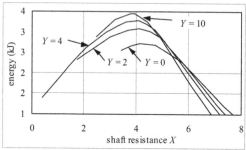

Figure 12. Dissipated energy per cycle

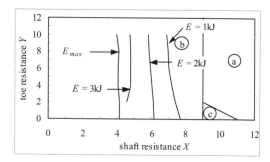

Figure 13. Equal energy curves

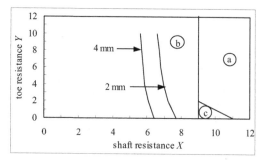

Figure 14. Equal amplitude curves

A lot of numerical tests were performed. We present here only the more significant results. Figure 12 shows the transmitted (or dissipated) energy per cycle versus $X$ and $Y$. The maximum increases with $Y$, but is not heavy dependent of $X$. To each maximum corespond an amplitude of displacement and a penetration speed. These are given in table 6. $X_{max}$ is the value of X corresponding to the maximum of energy.

Table 6. Maximum energy, amplitude, and penetration speed

| $Y$ | 0 | 2 | 4 | 10 |
|---|---|---|---|---|
| Maximum energy (kJ) | 2.72 | 3.08 | 3.28 | 3.44 |
| $X_{max}$ | 4.14 | 4.12 | 4.07 | 3.96 |
| Amplitude of displac. (mm) | 8.01 | 8.04 | 8.15 | 8.08 |
| Penetration speed (cm/s) | 22.8 | 14.5 | 10.7 | 6.08 |

The amplitudes of displacement corresponding to the maximum of the energy (see Table 6) do not depend on $Y$, and is about 78% of the "free" amplitude of displacement. On the other hand, the penetration speed is highly depending on this parameter $Y$ and increases when the maximum of energy decreases. Therefore the maximum of energy cannot be a satisfactory criterion for the penetration speed without other criteria. As additional information, the Figure 13 shows that the transmitted energy depends clearly little or none on $Y$.

### 3.5 Amplitude criterion

It is clear that energy is an important parameter for the penetrability. In our opinion, the amplitude is also an important parameter in order to overcome the resistance of the soil.

Figure 14 shows equal displacement amplitude for two cases : an amplitude peak to peak of 2 mm (10% the "free" amplitude), and an amplitude of 4 mm (20% of the "free" amplitude). For the second case, the shaft resistance $X$ is close to 6 and more or less no depending from the toe resistance $Y$. The curve corresponding to the first case is more or less close to the equal energy 1kJ and the second case to the equal energy 2kJ).

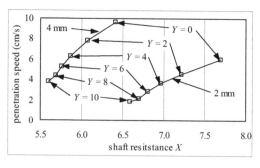

Figure 15. Penetration speed for equal amplitude

As shown on Figure 15, the penetration speed depends highly on the soil resistance for same amplitudes of displacement. Therefore a limit defined by amplitudes cannot be a satisfactory criterion for the penetration speed without other criteria.

## 3.6 Conclusion

The question of the prediction of the refusal depth remains open: with a plastic behaviour for the soil, penetration speed, amplitude and energy decrease to nil simultaneous. A criterion on energy or a criterion on amplitude of displacement are not sufficiently discriminating. The question can be solved by introducing an other definition of the refusal penetration: instead of a nil penetration speed, it is possible to consider a finite value for the final penetration speed. But which value is to be chosen for that limit?

## 4  QUAKE EFFECTS

This modeling of the soil used previously can be considered as too simplified. An elastoplastic behaviour is perhaps more adapted and can contribute to solve some problems. We will now analyse the effects of incorporating a quake in our model.

### 4.1  Effect on the penetration speed

At first, we will analyze the behaviour of the "system" for same values of the toe and shaft resistance as previously, and for some toe quakes $q_p$ and shaft quakes $q_l$.

Penetration speed $V_f$ and dissipated energy $E$ are given in table 7. It is clear that the introduction of quakes increases both $V_f$ and $E$ initially obtained with those initially obtained using the plastic model: the foreseeable refusal depth will be greater with quakes than without quakes.

Table 7. Comparison for $X = 7$ and $Y = 4$

| quakes | $v_f$ (cm/s) | E (kJ) |
|---|---|---|
| $q_l = q_p = 0$ | 3.22 | 1.16 |
| $q_l = q_p = 1mm$ | 7.22 | 6.05 |
| $q_l = q_p = 2mm$ | 6.31 | 6.44 |
| $2q_l = q_p = 2mm$ | 5.47 | 6.08 |

Figures 16 and 17 give more information on the effects of the quakes respectively on the displacements and on the velocity. Of course, with an elastoplastic law, we have not "stop and go" phases. On the other hand, the curves calculated for both values of the quakes (1 and 2mm) are close. An other numerical test (not given on these figures) for $2q_l = q_p = 2mm$ gives results very close to these obtained for $q_l = q_p = 2mm$: it seems that the behaviour depends more on the toe resistance quake than on the shaft resistance quake.

The evolution of both toe and shaft resistances over one period of the vibrator are given respectively on Figure 18 and 19.

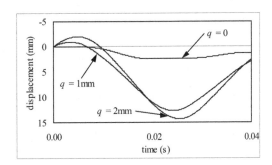

Figure 16. Displacements versus time and quake

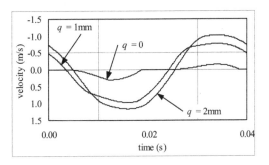

Figure 17. Velocities versus time and quake

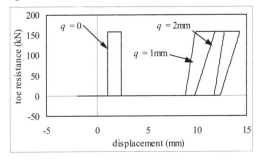

Figure 18. Toe resistance versus dis placement

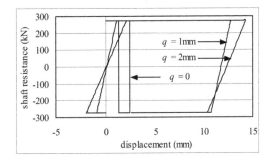

Figure 19. Shaft resistance versus displacement

### 4.2 *Refusal depth*

Resorting to quakes can perhaps solve the problems for the prediction of the refusal depth. Figure 20 shows an example ($Y = 4$) of the penetration speed versus shaft resistance (or depth). Without quake, the refusal depth was reached only at the asymptotic end of the curve. With quakes, we observe a sudden decrease of the penetration speed followed by numerical instability. We have no explanation for this numerical instability. Figure 20 gives also the curve corresponding to $2q_l = q_p = 2mm$ for which numerical instabilities do not occur. We have little explanation regarding that point.

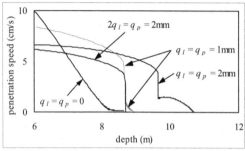

Figure 20. Penetration speed versus depth

If we choose this abrupt decrease before numerical instabilities for the refusal depth we obtain the results given on Figure 21.

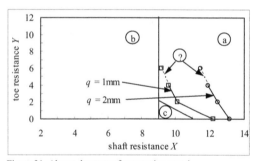

Figure 21. Abrupt decrease of penetration speed

### 4.3 *Conclusions*

The introduction of quakes does not give more final results concerning the prediction of the refusal depth. Without quake, the "stop and go" analysis and the energy analysis places the refusal depth on the left of the limit $X = 9$ in diagram $X - Y$. With quakes, the analyze places the refusal depth on the right of the same limit in the same diagram. It is difficult to conclude more precisely.

## 5 CONCLUSIONS AND OUTLOOK

The more important conclusions from this contribution are the following:

- a rigid body model assuming a rigid plastic behaviour of the soil around the shaft and the toe of the pile is sufficient to describe correctly penetration and extraction of medium piles (length smaller than 15 meters), at least for cohesionless soils. For other soils, it will be probably necessary to introduce quakes for both toe and shaft resistances. The software BRAXXUS is able to take into account these quakes;

- shaft and toe resistance during vibro-driving can be estimated from the static values. The cases analyzed here concern only homogeneous soils, but it is easy to introduce layers with various proprieties and characteristics;

- penetration speed is highly dependent on the quakes introduced in the modeling. That means it is very important to introduce realistic values of these parameters for a satisfactory penetration prediction. On the other hand, it seems that the behaviour depends more on the toe resistance quake than on the shaft resistance quake;

- more information concerning the dynamic behaviour around the pile is needed in order to progress with the prediction of the refusal depth. Typically, the frequency, amplitudes, accelerations effects on the toe and the shaft resistances are not well known. The planned laboratory tests at the Ecole Nationale des Ponts et Chaussées in France will certainly contribute to give a better knowledge on the exact dynamic behaviour of the soil around the pile.

## ACKNOWLEDGEMENT

The work presented in this paper was carried out in the framework of the French national project on vibro-driving. This project is managed by IREX and forms part of the operations of the RCG&U network. It is sponsored by the French Ministry of Public Works, which is acknowledged for its financial support.

## REFERENCES

Sieffert, J.-G. 2000. Prévision de la pénétrabilité. Méthode ENSAIS – Logiciel BRAXXUS. *Research report: IREX:* 29 pages.

Sieffert, J.-G. 2001. Vibrofonçage – Analyse d'un critère de refus. *Research report: IREX:* 11 pages.

Vié, D. 2001. Prévision du vibrofonçage – Modélisation simplifiée – Recueil et analyse des données. *Research report: IREX: documents 3 and 4:* 100 pages.

# Application of a hypoplastic constitutive law into a vibratory pile driving model

J-F. Vanden Berghe
THALES Geosolutions, Brussels, Belgium

A. Holeyman
Université catholique de Louvain, Louvain-la-Neuve, Belgium

ABSTRACT: This paper describes a model calculating the penetration speed of a pile during its vibratory driving and the vibrations induced around it. The model is called VIPERE, for VIbratory PEnetration REsistance.

The model actually implements hypoplastic constitutive behaviour into a geometric model suggested by Holeyman (1994). The model VIPERE considers the pile as a rigid body and simulates the soil by a 1D radial discretisation. The soil behaviour is assumed to be hypoplastic and modeled using the Bauer (1996) and Gudehus (1996) constitutive law. The penetration speed and the wave propagation around the pile are evaluated by integrating the equation of movement.

The paper describes the model by specifying the assumptions relative to the pile and the soil behaviour, the equations used to evaluate the soil resistance along the pile shaft and at the pile base, and finally the procedure of integrating the motion equations. Typical simulations of pile driving are presented and discussed.

## 1 INTRODUCTION

The VIPERE model calculates the penetration speed of a pile or sheet-pile at a given depth during a vibratory driving.

The model considers the pile as a rigid body and represents the soil with a cylindrical discretisation of rings. The interaction between these rings is described by a hypoplastic constitutive law (Fig. 1).

To calculate the displacement of the pile as well as the wave propagation around the vibrating profile, the model integrates the equation of motion step by step. The acceleration is calculated at each time step, based on the balance of the forces acting on the pile and on each ground element considered in the discretisation.

The forces acting on the pile are:
➤ the vibrating force induced by the vibrator (= $m_e.\omega^2.\sin(\omega t)$ where $m_e$ is the eccentric moment and $\omega$ is the angular frequency (= $2.\pi.\text{Freq}$)),
➤ the static weight placed above the vibrator and isolated from it by shock absorbers (= $M_{total} . g$),
➤ the friction resistance along the shaft of the pile $F_{shaft}$,
➤ the toe resistance $F_{toe}$ and
➤ the inertial force induced by the movement of the mass of the pile and the vibrator.

The forces acting on each ground element of the discretisation are, in addition to the inertial force, the forces of friction generated on the internal and external faces by the movement of the close elements.

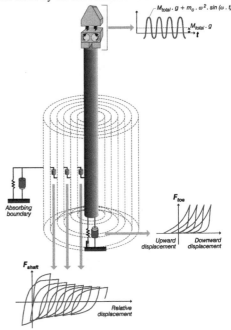

Fig. 1: Vibrodriving model.

During simulations of the vibratory driving of sheet-piles, the model does not take into account the influence of the sheet-pile wall formed by the other already installed sheet-piles neither the friction resistance generated at the interlock with the neighbour sheet-piles.

The scope of this paper is to describe the main assumptions considered in the VIPERE model and to show how the hypoplastic constitutive equation is implemented into the model. The integration procedure is also presented. Finally, the model performances are discussed through the analysis of simulation results.

## 2 GEOMETRIC MODEL

This model borrows the geometric configuration of the HYPERVIB II model developed by Holeyman (1994, 1996).

The geometric shape (Fig. 2 a) assumes the pile or sheet-pile and the surrounding soil have cylindrical symmetry. Since the shape of the pile is not necessary cylindrical, the equivalent radius of the pile is obtained from perimeter considerations. The soil is assumed to be a disk (Fig. 2 b) with a thickness that slightly increases in a linear way with the radius (Fig. 2). This increase tends to simulate the geomet-

rical damping provided by vertical diffusion of waves around the profile.

$$\Delta h = \alpha \cdot \Delta r \qquad (Eq. 1)$$

where r is radius and $\alpha$ is the coefficient of dispersion (= 0.03 – Holeyman, 1996)

In order to simulate the wave propagation, the soil is discretised into a set of concentric rings possessing individual mass and transmitting forces to their neighbouring ones (Fig. 2 c). The base resistance of each element is supposed to balance the gravity force. The vertical shear stress $\tau$ between two rings is calculated based on the relative displacement of each ring using the hypoplastic model (Gudehus, 1996; Bauer, 1996). The radial discretisation is characterised by the number of rings (Nr) and the maximum radius ($R_{max}$).

The boundary condition of the cylindrical discretisation is selected in such way that arriving waves are absorbed by that border (Novak, 1978). However, considering Novak's assumptions, this border will absorb the incident wave only if the wave induces a displacement of the soil element that stays in the linear elastic domain of the soil. In order to respect that condition, the number of soil elements and the maximum radius must be large enough to ensure the wave is damped enough at the boundary and is in the elastic domain.

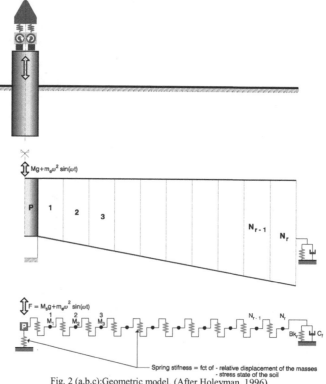

Fig. 2 (a,b,c):Geometric model. (After Holeyman, 1996)

# 3 CONSTITUTIVE LAW FOR SHAFT RESISTANCE

## 3.1 Assumptions

The model considers each soil element as purely sheared (Fig. 3). A shear displacement is imposed between the internal and external shaft of the soil element. Neither axial nor radial normal strains are permitted along these boundaries. The non-vanishing strain of the strain tensor is the shear strain $\gamma_{rz}$ ($\gamma_{rz} \neq 0$; $\varepsilon_r = \varepsilon_\theta = \varepsilon_z = 0$).

The soil is assumed to be fully saturated and the frequency of the vibrator is supposed high enough to prevent excess pore water dissipation during the vibratory driving. Therefore, the model considers the behaviour of the soil is undrained (i.e. no volume change; $\Delta e = (1+e).(\varepsilon_r+\varepsilon_\theta+\varepsilon_z) = 0$).

It is assumed that there is no variation of the stresses and strains along the z axis ($\partial(..)/\partial z = 0$). The radial normal stress and the shear stress do not change as a function of the depth.

Based on these assumptions, Fig. 4 shows the stresses and strains distribution at each soil element interface. The shear strain $\gamma_{rz}$ is calculated based on the relative displacements of these elements assuming a linear distribution of the displacement.

The resulting stress tensor $T_s$ and strain tensor $D_s$ can be written as follow[1]:

$$T_s = \begin{pmatrix} -\sigma_r' & 0 & \tau_{rz} \\ 0 & -\sigma_\theta' & 0 \\ \tau_{rz} & 0 & -\sigma_z' \end{pmatrix} \quad D_s = \begin{pmatrix} 0 & 0 & \gamma_{zr} \\ 0 & 0 & 0 \\ \gamma_{rz} & 0 & 0 \end{pmatrix} \text{ (Eq. 2)}$$

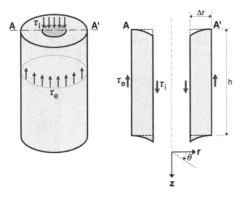

Fig. 3: Simple shearing in axisymetry

---

[1] For the hypoplastic constitutive equation, the compressive stresses and shorttening strains are defined negative in accordance to the convention in continuum mechanics. Only the scalar of the effective mean stress P' ( $= -$ tr(T$_s$).$\frac{1}{3}$ ) is defined positive in compression.

## 3.2 Hypoplastic model in cylindrical si shearing

The constitutive relationship used to calculate the shear stress along the shaft of the different elements is the hypoplastic model (Kolymbas, 2000, Gudehus, 1996, Bauer, 1996). This constitutive law evaluates the stress rate tensor $\dot{T}_s$ as a function of the current stress state tensor $T_s$ and the strain rate tensor $\dot{D}_s$ based on the following incremental equation:

$$\dot{T}_s = f_s .\left[a_1^2 . \dot{D}_s + \hat{T}_s . \text{tr}(\hat{T}_s . \dot{D}_s) + f_d . a_1 . (\hat{T}_s + \hat{T}_s^*). \|\dot{D}_s\|\right]$$

(Eq. 3)

$a_1$ is a dimensionless scalar depending of the friction angle $\varphi'$ of the soil, $\hat{T}_s = T_s/\text{tr}(T_s)$ is the granular stress ratio tensor, $\hat{T}_s^* = \hat{T}_s - \frac{1}{3}.1$ is the deviatoric part of $\hat{T}_s$, $f_s$ and $f_d$ are scalars depending of the mean stress P'$_s$ and the void ratio e.

Based on notations and assumptions of paragraph 3.1, equation 3 can be simplified and can be expressed in its incremental shape into the following components of the stress state tensor:

$$\Delta\tau_{rz} = f_s .|\Delta\gamma_{rz}|.\left[\left(a_1^2 + \frac{2.\tau_{rz}^2}{(\sigma_r'+\sigma_\theta'+\sigma_z')^2}\right).\text{sign}(\Delta\gamma_{rz}) \right.$$
$$\left. + f_d .\sqrt{2} .a_1 . \frac{2 .\tau_{rz}}{.(\sigma_r'+\sigma_\theta'+\sigma_z')}\right]$$

$$\Delta\sigma_r' = f_s .|\Delta\gamma_{rz}|.\left[\frac{2.\tau_{rz}.\sigma_r'}{(\sigma_r'+\sigma_\theta'+\sigma_z')^2}.\text{sign}(\Delta\gamma_{rz}) \right.$$
$$\left. + f_d .\sqrt{2} .a_1 . \frac{(5.\sigma_r'-\sigma_\theta'-\sigma_z')}{3.(\sigma_r'+\sigma_\theta'+\sigma_z'')}\right]$$

$$\Delta\sigma_\theta' = f_s .|\Delta\gamma_{rz}|.\left[\frac{2.\tau_{rz}.\sigma_\theta'}{(\sigma_r'+\sigma_\theta'+\sigma_z')^2}.\text{sign}(\Delta\gamma_{rz}) \right.$$
$$\left. + f_d .\sqrt{2} .a_1 . \frac{(5.\sigma_\theta'-\sigma_r'-\sigma_z')}{3.(\sigma_r'+\sigma_\theta'+\sigma_z')}\right]$$

$$\Delta\sigma_z' = f_s .|\Delta\gamma_{rz}|.\left[\frac{2.\tau_{rz}.\sigma_z'}{(\sigma_r'+\sigma_\theta'+\sigma_z')^2}.\text{sign}(\Delta\gamma_{rz}) \right.$$
$$\left. + f_d .\sqrt{2} .a_1 . \frac{(5.\sigma_z'-\sigma_r'-\sigma_\theta')}{3.(\sigma_r'+\sigma_\theta'+\sigma_z')}\right]$$

(Eq. 4)

The factor $f_d$ controls the transition to the critical state. The factor $f_s$ takes into account the increase of stiffness consecutive to an increase of the mean stress. These factors are function of the soil state defined by the void ratio e and the effective mean stress P' $(=(\sigma_r'+\sigma_\theta'+\sigma_z')/3)$ and of seven constitutive parameters ($e_{c0}$, $e_{i0}$, $e_{d0}$, $\alpha$, $\beta$, hs and n). These parameters define the relationship between the minimum, critical and maximum void ratios and the means stress.

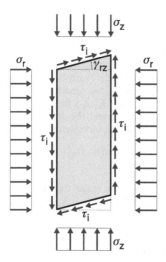

Fig. 4: Stress state assumed at the interface between 2 soil elements.

More detailed description of these parameters can be obtained in Bauer (1996), Gudehus (1996) or Vanden Berghe (2001).

# 4 CONSTITUTIVE LAW FOR TOE RESISTANCE

## 4.1 Assumptions

In order to calculate the toe resistance of the pile during the vibratory driving, the VIPERE model represents the soil under the pile by a cylinder (Fig. 5).

The section of the cylinder is equal to the section of the pile and its height is equal to 70% of the pile diameter. The soil at the pile base and the pile are supposed to stay permanently in contact. The model does not take into account the advent of gaps between the pile base and the soil as experimentally observed (Gudmani, 1997; Holeyman & al., 1998).

The determination of the height of the cylinder was suggested by the simplified semi-empirical equation calculating the immediate settlement of a foundation (Eq 5). This equation was developed based on the Boussinesq theory (Boussinesq – 1885), the theory of elasticity and experimental data. The influence coefficient $I_p$ depends principally of the shape of the foundation (Steinbrenner – 1934). As a first approximation, the value of this coefficient can be taken equal to 0.7 for a solid vibratory driven pile.

$$s = q.B.I_p.\frac{1-\nu^2}{E} \qquad \text{(Eq. 5)}$$

Where s is the foundation settlement, q is uniform contact pressure, B is the foundation width, $I_p$ is the influence coefficient, E is the soil stiffness and $\nu$ is the Poisson's ratio of soil

The behaviour of the soil cylinder at the pile toe is assumed to be hypoplastic and the element is supposed to be loaded in triaxial conditions (i.e. vertical stresses $\sigma_z$' and radial stresses $\sigma_r$' are principal).

As for the shaft resistance, the frequency of the cyclic loading is assumed to be high enough to consider that the excess pore pressure cannot be dissipated during the driving: the soil behaviour is assumed to be undrained ($\Delta e=0$). The pore pressure u is calculated based on the assumption that the total mean stress P ($=(\sigma_z+2.\sigma_r)/3$) of the considered soil element stays constant ($\Delta u=P'-P_0'$).

The resulting toe resistance $F_{toe}$ is calculated based on the total axial stress acting on the pile base using equation 6:

$$F_{toe} = (\sigma_z' + u) A_{pile} \qquad \text{(Eq. 6)}$$

where $\sigma_z$' is the effective vertical stress, u is the pore pressure and $A_{pile}$ is the pile section.

## 4.2 Hypoplastic model in triaxial shear

Similarly to the shaft resistance, the constitutive relationship used to calculate the stresses at the pile base is the hypoplastic model defined with Eq 3.

Based on the assumptions presented in the previous paragraph, the stress tensor and the strain rate tensor can be simplified as follow:

$$T_s = \begin{pmatrix} -\sigma_r' & 0 & 0 \\ 0 & -\sigma_r' & 0 \\ 0 & 0 & -\sigma_z' \end{pmatrix} \quad \dot{D}_s = \begin{pmatrix} -\frac{1}{2}\Delta\varepsilon_z & 0 & 0 \\ 0 & -\frac{1}{2}\Delta\varepsilon_z & 0 \\ 0 & 0 & \Delta\varepsilon_z \end{pmatrix}$$
$$\text{(Eq. 7)}$$

The vertical normal strain $\varepsilon_z$ is calculated by dividing the pile displacement $u_p$ by the height of the soil element considered under the pile ($=0.7.\varnothing_{pile}$).

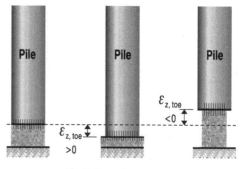

Fig. 5: Toe resistance model

Using these tensors, the stress rate at the pile base can be expressed as a function of the deviator q ($=\sigma_z'-\sigma_r'$) and the effective mean stress P' ($=(\sigma_z'+2.\sigma_r')/3$):

$$\Delta P' = f_s . |\Delta\varepsilon_1| . \frac{1}{3} . \left[ \frac{q}{3.P'} . sign(\Delta\varepsilon_1) + f_d . \sqrt{\frac{3}{2}} . a_1 \right]$$

$$\Delta q = f_s . |\Delta\varepsilon_1| . \left[ \left( \frac{3}{2} . a_1^2 + \left( \frac{q}{3.P'} \right)^2 \right) . sign(\Delta\varepsilon_1) \right.$$
$$\left. + f_d . \sqrt{6} . a_1 . \frac{q}{3.P'} \right]$$

(Eq. 8)

The density factor $f_d$ and stiffness factors $f_s$ are functions of only the effective mean stress P' and the void ratio e.

# 5 MOTION EQUATION AND INTEGRATION

## 5.1 Pile Equilibrium

Fig. 6 shows the different forces acting on the pile. The vibrator induces a force resulting from the gravity force and the cyclic force induced by the vibration of eccentric masses. The soil reaction is divided into the resistance along the shaft and the toe resistance. The shaft and toe resistances are calculated using the hypoplastic model based on the relative displacement between the pile and the soil elements and the stress state in these elements.

The acceleration of the pile results from the unbalance between these forces (Eq. 9)

$$\ddot{u}_p(t) = \frac{M_{tot}.g + me.\omega^2.sin(\omega.t) - 2.\pi.r_1.h_1.\tau_1(t) - F_{toe}}{M_{vib}}$$

(Eq. 9)

where $\ddot{u}_p(t)$ is the acceleration of the pile, $M_{tot}$ is the total mass of the vibrator and the pile, $M_{vib}$ is the vibrating mass consisting of the pile, the clamping device and the exciter block, me is the eccentric moment, $\omega$ is the angular frequency, $r_1$ is the equivalent radius of the pile, $h_1$ is the current pile penetration, $\tau_1$ is the shear stress at the interface between the pile and the soil and $F_{toe}$ is the base resistance

## 5.2 Soil Elements Equilibrium

Fig. 7 shows forces acting on each soil element modelling shaft resistance. The gravity force of the element is not taken into account. It is supposed to be balanced by the base resistance. The inter-ring reaction is calculated assuming an uniform distribution

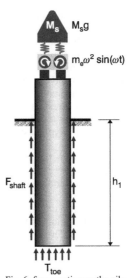

Fig. 6: forces acting on the pile

of the shear stress along the internal and external shaft of the soil element (Eq 10).

$$T_i = 2\pi r_i h_i \tau_i$$ 

(Eq. 10)

where $T_i$ is the inter-ring reaction between the elements i and i-1, $r_i$ is the radius of the interface between elements i and i-1, $h_i$ is the mean height of elements i and i-1, $\tau_i$ is the shear stress at the interface between elements i and i-1

Based on the inter-ring resistance, the acceleration $\ddot{u}(t)_i$ of each ring is calculated using equation 11. The displacement of the element is obtained by a double integration of this acceleration.

$$\ddot{u}(t)_i = \frac{(T_i - T_{i+1})}{M_i}$$

( Eq. 11)

where $\ddot{u}(t)_i$ is the acceleration of the elements i, $T_i$ is the inter ring reaction between the elements i and i-1, $T_{i+1}$ is the inter ring reaction between the elements i+1 and i, $M_i$ is the mass of the element i.

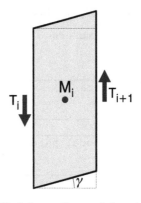

Fig. 7: forces acting on soil elements

The different steps in the calculation of the displacement of the soil elements i at the time step t+Δt can be summarised as follow:

➢ Based on the relative displacement between the considered element and its neighbours, calculation of the shear strain and the shear strain rate at each interface of the element;
➢ Based on the current stress state and the shear strain rate, calculation of the shear stress at each interface using the hypoplastic constitutive model;
➢ Calculation of the force acting an each interfaces by integration of the shear stress.
➢ Calculation of the acceleration of the considered element at the time t
➢ Calculation of the displacement of the element at the time t+Δt by integration of the motion equation using the central difference method.
➢ Return to the step 1

## 6 INTEGRATION PROCEDURE

Considering the motion equations of the pile and of each soil element, the system of Nr+1 motion equations of all the system can be developed. This system is not linear and cannot be solved by a direct inversion procedure. Indeed, the shear stress is calculated with the hypoplastic model that is expressed in an incremental shape. The model requires an explicit method to integrate the calculated acceleration.

The acceleration is evaluated as a function of the displacement using a central finite difference described as follow:

$$\ddot{u}_i(t) = \frac{u_i(t-\Delta t) - 2.u_i(t) + u_i(t+\Delta t)}{\Delta t^2}$$  ( Eq. 12)

The value of the displacements at the time t+Δt is calculated using Eq 12 where $\ddot{u}_i(t)$ is evaluated with Eq 9 and 11.

## 7 MODEL RESULTS

This paragraph presents the results calculated by the model VIPERE. These results are based on the input parameters described in Table 1 and on the modelling of the pile penetrating a homogenous layer down to 12m depth.

### 7.1 Pile penetration

#### 7.1.1 Shaft resistance

Along the shaft of the pile, the model VIPERE considers the soils condition is cylindrical simple sheared. The shear strain-shear stress (γ - τ) relationship is calculated using the hypoplastic constitutive law that takes into account the contractive and dilative behaviour of soil.

During the simulation of the vibratory driving, the $\tau_{rz}$ - $\gamma_{rz}$ hysteresis loops look like bananas (Fig. 8-a) similar to the shape observed during typical laboratory testing. That result is the consequence of the 2 phases of dilation and 2 phases of contraction that are observed during each cycle (Fig. 8-b). When the shear strain rate changes sign, the soil behaviour becomes contractive and the shear decreases rapidly towards 0 (section 1-2 on Fig. 8). While the behaviour is contractive (section 2 –3), the shear stress stays low. However, when the soil skeleton is no more able to follow the imposed shear strain without trying to increase its volume (point 3), the behaviour becomes dilative and the resulting shear stress increases significantly (section 3-4) until the direction of the shearing is inversed. These phenomena are also illustrated by the butterfly shape characterising the stress path in the P'-q plane (Fig. 8-c).

Table 1: Model parameters for simulations.

| Model Parameters | | | | | |
|---|---|---|---|---|---|
| Vibrator parameters | Eccentric Moment me = 46 kg.m<br>Frequency = 33 Hz<br>Vibrating Mass of Vibrator = 6700 kg<br>Stationary mass of Vibrator = 3500 kg | | | | |
| Pile Parameters | Pile Area = 167 cm²<br>Pile Perimeter = 288 cm (diameter = 92cm)<br>Pile Length = 12 m | | | | |
| Soil Parameters | φ' = 30° | | e = 0.60 | | |
| Hypoplastic model paramet | $e_{c0}$ = 0.88<br>$e_{d0}$ = 0.52<br>$e_{i0}$ = 1.21 | | hs = 200 Mpa<br>n = 0.35 | | α = 0.25<br>β = 1.10 |
| Integration Parameters | $R_{max}$ = 100 m<br>Number of soil elements = 100<br>Δt = $15.10^{-6}$ s | | | | |

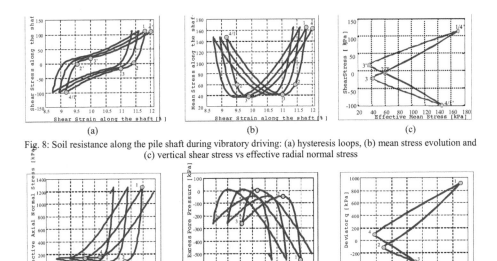

Fig. 8: Soil resistance along the pile shaft during vibratory driving: (a) hysteresis loops, (b) mean stress evolution and (c) vertical shear stress vs effective radial normal stress

Fig. 9: Soil resistance at the pile base during vibratory driving: (a) evolution of the effective normal axial stress, (b) evolution of the pore pressure, (c) stress path.

## 7.1.2 Toe resistance

The VIPERE model calculates toe resistance based on the total axial normal stress applied at the base of the pile. The effective axial normal stress $\sigma'_{z\,toe}$ (Fig. 9-a) is evaluated with the hypoplastic model assuming the pile displacement solicits the soil under the toe in a triaxial way. The pore pressure u (Fig. 9-b) is deduced from the calculated effective mean stress P' assuming that the total mean stress P stays constant around the base of pile. During each cycle, 2 phases of contraction and 2 phases of dilation are observed. When the pile moves downwards, the effective normal stress increases dramatically during the dilation phase (section 4-1 on Fig. 9-a). When the direction of pile displacement changes (point 1), the effective axial normal stresses decreases rapidly to a low constant value (section 1-2) and stays around that value during the following contraction and dilatation phases (section 2-3-4). In fact, when the pile moves upward, the soil behaviour becomes active: it is the lateral stress that pushes the soil towards the pile base.

The evolution of the toe resistance calculated by the VIPERE model is quite similar to the model proposed by Dierssen (1994). When the pile moves downward, the toe resistance stays small during a part of the downward displacement whereas it increases rapidly when a certain threshold is reached. In the VIPERE model, the threshold is crossing from the contractive behaviour to the dilative behaviour. In the Dierssen model, when the pile moves upward, the toe resistance decreases rapidly and is equal to 0. In the VIPERE model, the toe resistance is not equal to 0 but is very small depending of the soil.

## 7.1.3 Penetration speed of the pile

The evolution of the penetration speed deduced from the net penetration of the pile is plotted on Fig. 10.

Fig. 11 presents the evolution during 3 cycles of the different forces acting on the pile. The active force of the vibrator is a sinusoid lightly shifted upwards to take the static weight into account. The resisting force is less regular due principally to the different phases of dilatancy and contraction of the soil. The curve of the resisting force looks symmetric because the pile cross sectional area chosen for this simulation is small comparing to the area of the pile shaft: the soil resistance is principally applied along the pile shaft and the toe resistance is very small (around 2.5% of the shaft resistance).

The acceleration resulting of the unbalance between acting and resisting forces is integrated (Fig. 12) to calculate the vertical velocity and displacement of the pile.

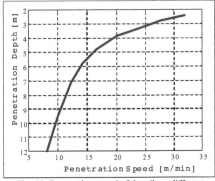

Fig. 10: Penetration speed of the pile at different depths.

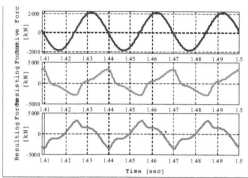 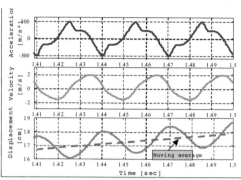

Fig. 11: Comparison of the forces acting on the pile (active
force = $M_{tot}.g+me.\omega^2.\sin(\omega t)$; resisting force = $-(F_{shaft}+F_{toe})$)

Fig. 12: Comparison of the acceleration, the velocity and the displacement of the pile

## 8 CONCLUSIONS

This paper has presented the VIPERE model calculating the penetration speed of a pile during its vibratory driving. More detailed description of this model can be found in Vanden Berghe (2001).

The model implements hypoplastic constitutive behaviour into a geometric model suggested by Holeyman (1994). The VIPERE model considers the pile as a rigid body and simulates the soil by a 1D radial discretisation. The soil behaviour is assumed to be hypoplastic and modelled using the Bauer (1996) and Gudehus (1996) constitutive law. The penetration speed are evaluated by integrating the equation of movement.

## 9 ACKNOLEGMENT

The authors thank the Fonds National de la Recherche Scientifique, the European Commission (Marie Curie Research Training Grant) and the Université catholique de Louvain who funded the research.

## 10 REFERENCES

BAUER, E. (1996). "Calibration of a Comprehensive Hypoplastic Model for Granular Materials." Soils and Foundations, Vol. 36, N°1, pp.13-26.

BAUER, E. (2000). "Conditions for embedding Casagrande's critical states into hypoplasticity." Mechanics of Cohesive-Frictional Materials; Vol. 5, pp. 125-148.

BAUER, E. and HERLE, I. (2000). "Stationary states in hypoplasticity. In Constitutive Modelling of Granular Materials" (Ed. KOLYMBAS D.), Springer, pp. 167-192.

BOUSSINESQ (1885). "Application des potentiels à l'étude de l'équilibre et du mouvement des solides élastiques. " Paris.

CUDMANI, R. (1997). "Ein bodenmechanisches Modell des Vibrationsrammens für nichbindige Böden. " Workshop Vibrationsrammen, Universiteit Karlsruhe.

DIERSSEN, GUILLERMO, (1994), "Ein Bodenmechanisches Modell zur Beschreibung des Vibrationsrammens in körningen Böden", Doctoral Thesis, University of Karlsruhe, Germany

GUDEHUS, G. (1996). "A comprehensive constitutive equation for granular materials. " Soils and Foundations, Vol. 36, N° 1, 1-12.

HOLEYMAN, A. & LEGRAND, C. (1994). "Soil Modelling for pile vibratory driving". International Conference on Design and Construction of Deep Foundations, Vol. 2, pp 1165-1178, Orlando, U.S.A., 1994.

HOLEYMAN, A., LEGRAND, C., and VAN ROMPAEY, D., (1996). "A Method to predict the driveability of vibratory driven piles". Proceedings of the 3rd International Conference on the Application of Stress-Wave Theory to Piles, pp 1101-1112, Orlando, U.S.A., 1996.

HOLEYMAN, A. and LEGRAND, C. (1997). "Soil-structure interaction during pile vibratory driving." Proceedings of the XIV[th] International Conference on Soil Mechanics and Foundation Engineering, September 6-12, 1997, Hamburg, Germany, pp. 817-822.

HOLEYMAN, A., VANDEN BERGHE, J.-F., and DE COCK, S. (1999). "Toe resistance during pile vibratory penetration. " Proceedings of XII[th] European Conference on Soil Mechanics and Geotechnical Engineering, June 1999, Amsterdam, Netherlands, pp. 769-776.

HOLEYMAN, A. (2000). "Vibratory Driving Analysis – Keynote Lecture. " Proceedings of the VI[th] International Conference on the Application of Stress-Wave Theory to Piles, September 11, 12 & 13, 2000, Sao Paulo, Brazil.

KOLYMBAS, D.(1985)°. "A generalized hypoelastic constitutive law." Proceedings of the XI[th] International Conference Soil Mechanics and Foundation Engineering, Vol. 5, A.A. Balkema, Rotterdam.

KOLYMBAS, D. (2000). "Introduction to Hypoplasticity. " A.A. Balkema, Rotterdam.

NOVAK, M., NOGAMI, T., and ABOUL-ELLA, F. (1978). "Dynamic Soil Reactions for Plane Strain Case", Journal of Engineering Mechanics Division, ASCE, 104(4), 953-959.

STEINBRENNER, W. (1934). "Tafeln zur Setzungsberechnung". Die Strasse, Vol. 1, Oct. 1934.

VANDEN BERGHE, J.-F., and HOLEYMAN, A. (1997). "Comparison of two models to evaluate the behaviour of a vibratory driven sheet-pile. " Proceedings of the XI[th] Young Geotechnical Engineers'conference, September 24-27, 1997, Madrid, Spain, pp. 60-72.

VANDEN BERGHE J-F, HOLEYMAN A., DYVIK R., Comparison and modelling of sand behavior under cyclic direct simple shear and cyclic triaxial testing, published in the proceedings of the Fourth International Conference on Recent Advances in Geotechnical Earthquake Engineering, San Diego, March 2001.

VANDEN BERGHE, J-F, (2001) "Sand Strength degradation within the framework of pile vibratory driving", doctoral thesis, Unversité catholique de Louvain, Belgium, 360pages.

# Simple model for prediction of vibratory driving and experimental data analysis of vibro-driven probes an sheetpiles

D. Vié
*CEBTP Saint Rémy les Chevreuse, France*

ABSTRACT: During the first part of the 'Vibrofonçage' French national project, experimental data has been given by the CSTC from the HYPERVIB project and by ARBED from the SIPDIS project. Two type of models has been proposed. At first, a simple model assuming the pile is a rigid body, gives an easy graphic analysis and simple criteria for refusal prediction. However, more sophisticated model are necessary for experimental data analysis. Based on classical solutions for dynamic test of piles, a 1D finite elements model is described. Characteristics of vibro driving analysis are pointed out. Examples are given from SIPDIS data. General conclusions give advice on experimental data analysis and evaluation of dynamic soil resistance.

## 1 INTRODUCTION

Two years ago, at the beginning of the French national project, CEBTP did not have experience of vibro-driving analysis, but had worked on dynamic test of piles. Although the fact that some authors have consider that vibro-driving analysis may be based on a simple rigid body model appeared rather surprising, we thought that these simple models may well give a rapid evaluation of refusal. The first part of this paper presents a solution for this simple analysis and gives indications on its use for evaluation of the mean velocity of the pile. Refusal criteria are discussed.

However, it also appeared that this simple models may not be very useful for experimental data analysis and that, for this purpose at least, better solution would be an adaptation of software devoted to pile dynamics analysis. In this context, we pointed out at first that this stress-wave analysis is often based on a high frequency approximation of the soil dynamic resistance, and so that this approximation must be used carefully.

A proper model would not be very useful without a good evaluation of soil parameters. Some authors have given evaluation of soil dynamic properties during vibro driving. But at the beginning of our project, our concern was to precise if soil parameters may be deduced from experimental datas. From both high strain or low strain dynamic test of piles, following ASTM standards denominations, this evaluation is possible by the wave equation analysis. During cyclic loading, soil reaction analysis may not be undertaken in the same way.

## 2 SIMPLE RIGID BODY MODEL

### 2.1 *Symbols and notations*

| | |
|---|---|
| $\Gamma, V, D$ | Acceleration, velocity and displacement |
| $M_{vib}$ | Dynamic mass including vibrator and pile |
| $F_{stc}$ | Static force (static weight of vibrator) |
| $F_{pte}, F_{lat}$ | Soil toe ands lateral resistance (time dependent) |
| $F_{rpt}, F_{rlt}$ | Limit values of soil toe and lateral resistance |
| $\omega, \varphi$ | Pulsation and phase between force and acceleration |
| $F_{exc}$ | Dynamic force of vibrator |

In the followings force and displacement are considered to be positives downward.

### 2.2 *Principle*

The most simple way to analyse the motion of a vibrodrived pile or sheetpile is to consider it as a rigid body mass. Soil resistance is assumed to be rigid plastic for shaft friction as for toe resistance. Examples of this simple approach may be seen in HYPERVIB I (A. Holeyman) and BRAAXUS (J.G. Sieffert).

The general equation of the pile motion is expressed by :

$$M_{vib}\Gamma = F_{stc} - F_{pte} - F_{lat} + F_{exc}\sin(\omega t - \varphi) \qquad (1)$$

The phase constant $\varphi$ is not a priori known and we may ask if there is only a solution independent on the initial conditions. This may be a matter of some importance for practical applications, where the pile motion is computed by applying some period of an harmonic force at the pile top.

The question is then to precise if the solution obtained depends or not on the choice of the phase φ. A detailed analysis of the pile motion shows how this question may be answered and gives indications on the domain of soil resistance values (shaft friction and toe resistance) where vibro-driving is possible.

In a rigid body analysis, there are only four possible situations :

(a) – upward motion
$F_{lat}=+F_{rlt}$ and $F_{pte}=0$ ;
(b) – downward motion without toe contact
$F_{lat}=-F_{rlt}$ and $F_{pte}=0$ ;
(c) – downward motion with toe contact
$F_{lat}=-F_{rlt}$ and $F_{pte}=F_{rpt}$ ;
(d) - no motion at all.

In each case, acceleration is determined by the equation (1). Considering the continuity of the velocity, we can see that upward motion and downward motion with toe contact are determined by the condition V=0. During the upward motion, the pile toe is not in contact with the soil and we suppose that downward motion will be without toe contact until total displacement from the beginning of the upward motion is equal to zero. The end of downward motion without toe contact is determined by the condition D=0.

## 2.3 *Reduced equations*

Values of acceleration, velocity may be expressed from the followings 'reduced' functions

$$f(x) = \lambda + \sin(x - \varphi) \text{ and } g(x) = g_0 + \int_{x_0}^{x} f(x) \qquad (2)$$

Then we may consider with $X = \omega t$

$$\Gamma(t) = \frac{F_{exc}}{M_{vib}} f(\omega t) \text{ and } V(t) = \frac{F_{exc}}{\omega M_{vib}} g(\omega t) \qquad (3)$$

giving to λ the proper value depending of the actual motion phase.

Another expression can be deduced for displacement.

So we can see that solutions for f and g functions depend only on the three followings parameters :

$$\alpha = \frac{F_{stc}}{F_{exc}} \quad \beta = \frac{F_{rpt}}{F_{exc}} \quad \gamma = \frac{F_{rlt}}{F_{exc}} \qquad (4)$$

Assuming that there is a point where V=0 the solution is entirely determined by the phase constant φ :

- the end of an upward motion or of a downward motion with toe contact is determined by the condition (V=0), and the equation to solve may be expressed as

$$ax + b + \sin(x - \varphi) = 0. \qquad (5)$$

- the end of a downward motion without toe contact is determined by the condition (D=0), and the equation to solve may be expressed as

$$ax^2 + bx + c + \sin(x - \varphi) = 0.$$

These equations are easily solved by iterations.

If there is no point where V=0, the motion is always in the same direction and the values of soil toe and lateral resistance are both constants.

## 2.4 *Vibro-driving domain*

In practical cases, value of $M_{vib}$, $F_{exc}$ and $F_{stc}$ are generally quite constant, but $F_{rpt}$ and $F_{rlt}$ are varying with the depth. Prediction have to give an evaluation at least for mean velocity.

At first, we may point out that if α+γ>1, that is to say $F_{stc}+F_{lat}>F_{exc}$, the upward motion will not be possible.

In the same way, if β+γ>α+1, that is to say $F_{rpt}+F_{rlt}>F_{stc}+F_{exc}$, the forces applied on the pile will never exceed the soil resistance. However, if an upward motion is possible, a downward motion may occur, at first without contact between pile toe and the soil and then after contact because of the velocity continuity.

These points may be resumed in the simple diagram of the figure 1, with the followings notations :

- (1) no motion at all
- (2) no upward motion
- (3) alternatively upward and downward motion or no motion.

In this last case, for γ<1-α, the motion is possible independently of the toe resistance, but as it will be shown later, the mean velocity decrease rapidly as the toe resistance increase.

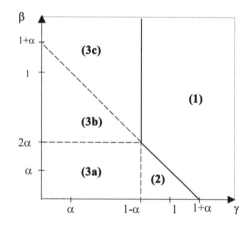

Figure 1 – Vibrodriving domain

The possibility of driving while applied forces never exceed soil resistance must be related to the assumption of a loss of contact at the pile toe during an upward motion.

Assuming that the applied force is harmonic, A. Holeyman and H. Gonin analyses the mean velocity as a balance between two phases of respectively upward and downward acceleration, following figure 2. Consequently, driving is only possible when $\beta<2\alpha$, that is to say $F_{rpt}<2F_{stc}$.

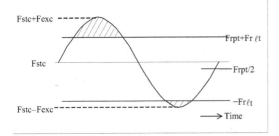

Figure 2 – Vibrodriving according Hypervib I

### 2.5 *Search of a periodic solution*

As mentioned in § 2.3, the pile motion is determined by a phase constant. We also could consider that the velocity may be determined for a period [0,T] if the velocity is known for t=0.

Let us consider

$$\theta_1 = a\sin(\alpha+\gamma)$$
$$\theta_3 = a\sin(\alpha-\beta-\gamma)$$

It appears that the motion at t=0 may be described by the following table

| $\theta_1$ | $\pi-\theta_1$ | $\pi-\theta_3$ | | $\theta_3+2\pi$ | $\theta_1+2\pi$ |
|---|---|---|---|---|---|
| (a) | (b) | | (c) | | (d) |

(a) Beginning of upward motion for $\omega t=0$ ;
(b) Beginning of upward motion for $\omega t=\varphi+\theta_1-\pi$ ;
(c) Beginning of downward motion for $\omega t=0$ ;
(d) Beginning of upward motion for $\omega t=\varphi-\theta_3$

The figure 3 below shows the evolution of the velocity at the end of the period (t=T) as a function of the phase constant $\varphi$ for $\varphi\in[\theta_1,\pi-\theta_1]$, that is to say solutions where there is an upward motion phase. Solutions without upward motion should also be searched for $\varphi\in[\pi-\theta_3,\theta_3+2\pi]$.

A discontinuity of the studied function correspond to the point where the contact between the pile toe and the soil append just at the moment where applied force becomes smaller than soil resistance.

Figure 4 shows acceleration and velocity deduced from the value of phase constant, and as a result, the velocity is V=0 at the end of the period.

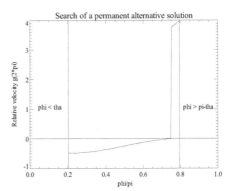

Figure 3 – Influence of the phase constant

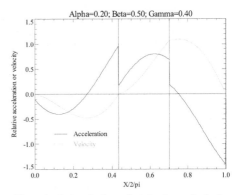

Figure 4 – Reduced values of acceleration and velocity

### 2.6 *Typical evaluation of mean velocity*

Figure 5 shows for $\alpha=0.10$ the evaluation of the relative mean velocity, that is to say the ratio between mean velocity and $V_{lib}=F_{exc}/\omega M_{vib}$, which would be the velocity of the pile without contact with the soil.

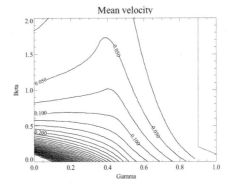

Figure 5 – Relative mean velocity for $\alpha=0.10$

Such graphic is easily drawn for each value of $\alpha$ parameter.

We may point out that :
- no periodic solution is found for $\beta+\gamma<\alpha$, this case corresponding to a fall of the pile under average acceleration ;
- the mean velocity decrease rapidly when $\beta$ or $\gamma$ increase.

Consequently, defining refusal condition will not be very easy : when mean velocity becomes very small it may be seen that this simple model is not very accurate : the dynamic amplitude of displacement or velocity does not tend to zero.

Nevertheless, the evaluation of mean velocity is very simple and the model has shown the necessity to consider the initial conditions in the search of solution : this will remain important when using more sophisticated models, which currently does not evaluate the velocity step by step all over the depth, but take some periods at each depth step. We must keep in mind that the number of period must be sufficient to obtain a periodic solution and that it may be not possible especially for small depth when soil resistance is not sufficient.

## 3  1 D FINITE ELEMENTS MODEL

Mechanic wave propagation in a pile is generally considered as a 1D problem, and lateral soil friction is supposed to be a sum of two terms, one being proportional to pile displacement and the second to pile velocity.

A brief glance at wave propagation in layered media gives useful information about this approximation and its validity both in the time domain and in the frequency domain.

Considering that radial displacement may be neglected before displacement along the z-axis, the general wave propagation is following the well known Bessel equation:

$$\rho_s \frac{\partial^2 D}{\partial t^2} = E_s \frac{\partial^2 D}{\partial z^2} + G_s \left( \frac{\partial^2 D}{\partial r^2} + \frac{1}{r} \frac{\partial D}{\partial r} \right) \tag{6}$$

where :  $\rho_s, E_s$ and $G_s$ are respectively the soil density, elastic Young and shear modulus,

D is the displacement along the z-axis,

R is the distance to the pile axis

t the time.

Solution of this equation may be obtained separately inside the pile and in the soil around the pile.

### 3.1  *Lateral friction in elastic conditions*

Considering that soil velocity is small before pile velocity, according to the following schedule, it appears that propagation around the pile is mainly due to horizontal shear wave propagation.

Analytical solution for the shear stress soil friction is the following

$$R(\omega) = 2\pi G_s x \frac{H_1(x)}{H_0(x)} D(\omega) \tag{7}$$

where  $x = \dfrac{r\omega}{\beta}$  and $H_n(x)$ is the Hankel function of the $n^{th}$ order.

Figure 6 shows the variations of the function $y=x(H_1(x) / H_0(x))$ and suggest that for high frequencies, that is to say when $x \gg 1$, a simple linear approximation may be used. It must be quoted that this expression is not defined for $x=0$, and must not be used to relate dynamic and static properties.

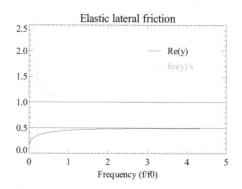

Figure 6 – Soil friction as a function of the frequency

A typical case is a pile diameter $\phi=0.50$ m and a soil shear wave velocity $\beta=500$m/s. The linear approximation may be used for $\omega>2000$ rad/s that is to say f>318 Hz. This assumption is correct for low strain dynamic test, where frequency range is about from 0 to 2 kHz, but less accurate for high strain dynamic test where frequencies are generally less than 1 kHz, and may be not accurate at all for vibrodriving frequencies which are much smaller, the fundamental mode being typically about 30 Hz..

## 3.2 Plasticity

For plasticity, SIMBAT method consider that, beyond shear pile/soil interface resistance, the shear stress is the sum of a term corresponding to the ultimate resistance and a term of viscosity due to local sliding between pile and soil.

$$\tau = \varepsilon \tau_l + \tau_g \left( V_p - V_s \right) \qquad (8)$$

where $\varepsilon = \text{sgn}(V_p - V_s)$

The viscous component $\tau_g$ is supposed to depend on the pile material and on the soil characteristics.

Identifying the dynamic shear stress with the precedent expression, we now obtain the expression of the total shear stress at the pile/soil interface

$$\tau = \varepsilon \tau_l + \frac{G_s}{\beta} k_g \Delta_1 V_p \qquad (9)$$

where $k_g = \dfrac{\beta \tau_g}{G_s + \beta \tau_g}$ and

$$\Delta_1(x) = \frac{H_1(x)}{(1 - k_g) H_1(x) + j k_g H_0(x)} \qquad (10)$$

As before, the use of the Bessel function approximation for large values of X let us give the general form of the dynamic shear stress between pile and soil around it.

$$R_d(\omega) = 2 \pi r \tau_d = \varepsilon \tau_l + k_g \pi G_s \left[ k_g D(\omega) + \frac{2r}{\beta} V(\omega) \right] \qquad (11)$$

The elastic values are obtained with $\varepsilon = 0$ et $k_g = 1$.

## 3.3 Frequency and time domain analysis

Since we have seen that, at least for the elastic case, the above equation have no solution for $\omega=0$, special attention must be accorded to their time domain signification.

Inverse Fourier transform of these function is not possible with a FFT algorithm, but may however be calculated with care of the low frequencies analysis. Figure 7 give the solution obtained for the function $\Delta_2 = (\Delta_1 - 1)/k_g$ in the time domain.

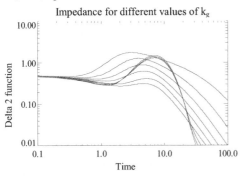

Figure 7 – Impedance function in the time domain

The general expression for the shear stress is therefore

$$\tau(t) = \varepsilon \tau_l + k_g \frac{G_s}{\beta} V(t) + k_g^2 \frac{G_s}{r} \int_0^\infty \delta \left( \frac{\beta u}{r} \right) V(t - u) du \qquad (12)$$

The limit of the delta function for small values of $\beta t/r$ is always $\delta = 0.5$ corresponding to the high frequency approximation. But we can see that this approximation is not valid as soon as $\beta t/r \geq 1$, that is to say, in the same case presented above t > 0.5 ms.

Typically for a frequency of 30 Hz, one period is about 30 ms and we have seen that some consecutive periods must be considered.

## 4 SIPDIS EXPERIMENTAL DATA ANALYSIS

At the beginning of our project, PROFILARBED provide us experimental data results from the SIPDIS project. This paper does not present a global analysis of these data but give information about what could be a method for this analysis and a first glance at some results. All data concerned here are from Hamburg site.

### 4.1 Geotechnical information

The Hamburg test site correspond to rather homogeneous soil conditions : below a rather dense –surface layer (about 10-20 MPa), consisting of sandy material down to about 6 m depth, follows an about 4 m thick layer with low cone resistance (0.5 – 1.5 MPa) with a high friction ratio. At almost exactly 10 m, the dense sand deposite starts, with a cone resistance of 15 – 30 MPa.

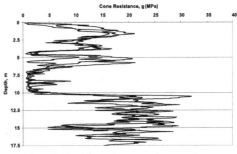

Figure 8 : Hamburg test - cone resistance values.

### 4.2 Piles tested

Numerous tests have been performed, including sheet piles type LX 16, LX 32, AZ26 and AZ 48, single or doubles, vibro-driving or vibro-jetting.

A special probe developed for the SIPDIS project has also been used. This probe is a pile equipped with extensive instrumentation : strain gauges and acceleration as well as temperature sensors. This probe consist of waterproof measuring elements as well as tubular sections. The modules are based on a steel section of 114.3 mm outside diameter and 98.3 mm inside. Three instrument modules are 0.75 m long, one at the top the second at the mid and the third at the bottom of the probe.

Figure 9 : SIPDIS probe.

### 4.3 *Instrumentation*

Until now, following measurement have been analysed :
- frequency ;
- dynamic force of the vibrator ;
- depth ;
- strain gauges ;
- acceleration on sheet piles.

Other information are available, as acceleration on the vibrator, velocity on the soil surface or at some depth, etc...

All sensors are connected to an acquisition device, and measurement are done with a frequency of 2 kHz for 0.5 s that is to say 1000 points every 2,75 s.

### 4.4 *Depth measurement*

As it is well known, acceleration sensor doe not give information about velocity and displacement. The first point was the evaluation of the mean velocity

and so the first step of experimental analysis are the depth measurements.

A least square approximation of depth measurement using $2^{nd}$ degree polynomial which also give an evaluation of the mean acceleration.

At the beginning of the test, we can find that the mean acceleration is not neglectful. It may be alternatively positive or negative, depending on the driver : until the soil resistance is sufficient, the pile must be hold back.

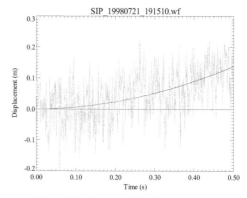

Figure 10 : adjustment of depth measurement

The second step is to have a general view of the test, generally :
- a first phase where the pile is hanging without contact with the soil ;
- the vibrodriving sequence, more or less complicated following experimental hasards ;
- the extraction sequence ;
- a last hanging phase.

Following figure 11 gives an example from test #05 that is to say a single LX 16 sheet pile 15 m length.

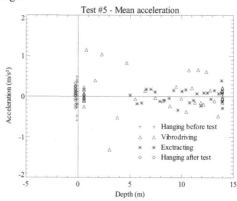

Figure 11 : mean acceleration values

In this case, mean acceleration are not neglectful under 5 m depth and also at the end of test, probably when the driver let all the vibrator mass push the sheet pile.

Then the mean velocity values shows also that until a 5 m depth was reached, mean velocity is rather constant, indicating that the sheet pile is hold back by the driver.

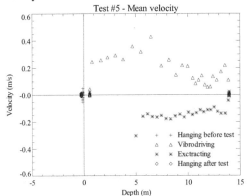

Figure 12 : mean velocity values

### 4.5 *Frequency adjustment*

Some difference have been observed between the frequency measured and the periodicity deduced from acceleration and force measurement. So, for each data set, a general least square approximation of the frequency has been searched.

The period T is supposed to be known, and we have a data set (ti,Yi) which generally covers about 15 periods. The adjustment is made with two steps :
-   a linear interpolation for tj values corresponding to an exact division of T and calculation of Xj=tj modulo T ;
-   calculation of Yj and his standard deviation for each value of Xj.

The value of T which give the minimal value of the total standard deviation for all the sensor considered is supposed to be the exact mean period for the data set.

An example of adjusted strain measurement is given below. The mean value has been stored for each data set to be used for further analysis and comparison with models.

In every case, strain and acceleration are found to be periodic, but some variations occurs especially when pile is hanging without contact with the soil. The following example shows a strain adjustment at the final stage of depth penetration.

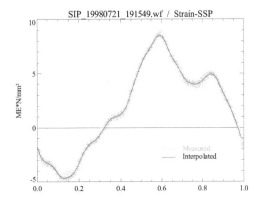

Figure 13 : periodicity adjustment of strain measurement

Following figure 14 shows difference between measured and adjusted frequency.

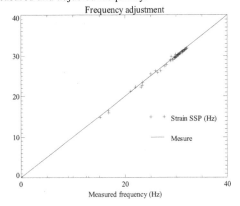

Figure 14 : measured and adjusted frequency

### 4.6 *Evolution of dynamic force*

After depth and frequency measurement, dynamic force may be evaluated from the 'UBL' data which give the angle between the two rotating mass of the vibrator.

The maximum value corresponding to $\delta=90°$, the dynamic force is

$$F_{dyn} = \omega^2 F_{max} \sin(\delta)$$

Following figure 15 shows the values for the same test #05 corresponding to a single LX 16 sheet pile. Dynamic force is quite constant all over the vibro driving and then during the extraction.

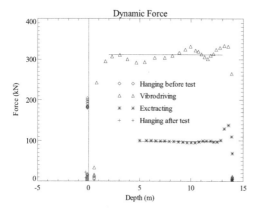

Figure 15 : dynamic force test #05

### 4.7 *Sheet pile hanging*

Measurement during the pile is hanging without contact with the soil is interesting because it gives direct information on the reliability of the sensors.

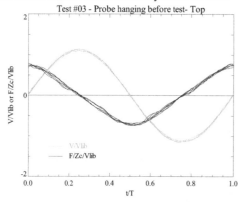

Figure 16 : relative force and velocity - test #03

Let us remember that in this case, force and velocity are related as follow

$$F = jZ_c Vtg\left(\frac{\omega l}{c}\right)$$

For a frequency about 30 Hz, and a pile 15 m length, we have $tg\left(\frac{\omega l}{c}\right) \approx 0.60$.

The velocity is harmonic and equal to the 'free velocity' Vlib in a $\pi/2$ phase with the force.

For the probe test #03 et #04 these relations are verified, as well for top, mid and bottom measurement.

### 4.8 *Strain mean values*

For test on probes, strain measurement are given in kN directly. These measurements included important offset values as shown below.

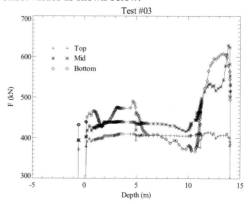

Figure 17 : force mean values - test #03

For the probe, the reference values are the followings:
- self weight about 3 kN ;
- total weight of vibrator 25 kN ;
- toe force for qc=10 MPa about 100 kN.

So we have to consider an offset of about 400 kN for all sensors.

It seems that :
- top mean strain is rather constant, varying from the self weight of the probe (traction) to the self weight of the vibrator (compression) ;
- bottom mean strain is varying with the depth following the toe resistance ;
- mid mean strain is varying mainly at the end of the penetration.

This mean value of strain must be used carefully. In many cases, the variation with depth are rather difficult to understand. Finally, the analysis of dynamic values for the pile hanging before test is the best control.

### 4.9 *Strain and acceleration dynamic values*

For a first glance at dynamic values, we have used a comparison between force and velocity, and for practical reasons used Vdyn/Vlib and Fdyn/Zc/Vlib, wich are adimensional values which may be directly compared and are introduced in what is called the dynamic resistance for dynamic test of piles.

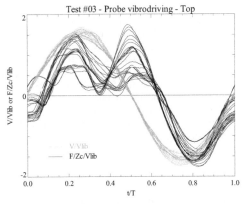

Test #03 - Probe vibrodriving - Top

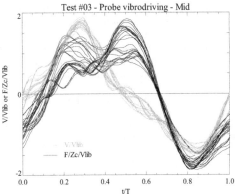

Test #03 - Probe vibrodriving - Mid

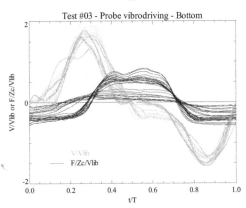

Test #03 - Probe vibrodriving - Bottom

Figure 18/20 : relative dynamic force and velocity test #03

The comparison is done for top, mid and bottom measurement, and includes every time step from 5 m depth to 13 m depth to see the evolution of force and velocity all over the penetration.

At the top, surprisingly, velocity is practically harmonic and more important than the dynamic 'free' velocity. The dynamic force is far more complex including high frequencies values. This high frequencies terms are depth dependent.

At the bottom, force is clearly rigid plastic, the mean value is zero because we have only the dynamic term, but the mean value has been shown before. Velocity shows alternatively downward and upward motion with steps of rather small values, as it has been discussed in the first part of this paper. Maximum values of velocity are also much higher than 'free' dynamic velocity.

At the mid pile, neither force or velocity are harmonic, but force is less complex at the top and velocity more regular than to the bottom.

### 4.10 Other information on test #03

Because information about dynamic force, mean acceleration and velocity §4.4 and §4.6 have been given for a single LX 16 sheet pile (test #05) following figure 21 give the evolution of the 'free' acceleration for the test #03.

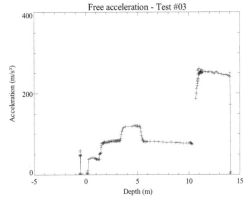

Figure 21 : free acceleration test #03

The free acceleration is rather constant up to 10 m depth (except from 4 m to 5 m depth) about 80 m/s², but increase up to 250 m/s² at the end of the test.

### 4.11 Dynamic acceleration and force ratio

It has been quoted that the dynamic acceleration (or velocity) may be greater than the 'free' velocity, that is to say the velocity of the pile without soil contact.

Analysis of the ratio between dynamic acceleration at the top of the pile and free acceleration is then interesting. An example is given below (figures 22 and 23) for test #04, wich is a probe, as test #03, but wich include measurement during extraction.

The ratio is increasing from about 1.1 when the pile is hanging before test up to 1.5 at the beginning of vibro driving, rather constant during driving, and decreasing slowly during extraction to return to 1.1 at the end of extraction.

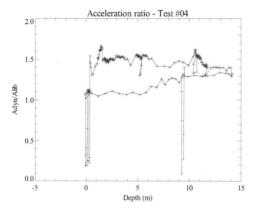

Figure 22 : acceleration ratio during vibro-driving
and extraction

In the same way, although the force at top is not harmonic, analysis of a dynamic force ratio is shown below.

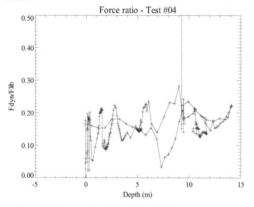

Figure 23 : force ratio during vibro-driving and extraction

Amplitude of the dynamic force (first mode) at the top of the pile is much smaller than the nominal force of the vibrator, the difference is mainly due to the 'dynamic' mass of vibrator.

## 5 CONCLUSIONS

### 5.1 *Global depth vs. time analysis*

Simple rules and evaluation of mean velocity has been proposed from a simple rigid body model. Results have been compared with more complex models but not remember here.

The main points remain :
- the choice of dynamic soil resistance, which need further experimental data analysis ;
- the need to search permanent motion, which occurs only when the soil resistance is sufficient, after a few meters depth.

Evaluation of the mean velocity is then accurate only when the soil resistance is sufficient : in the case of SIPDIS Hamburg test, we can say that the minimum depth is about 5 m. Under this depth, mean velocity depends mainly on the driver. A global depth/time analysis is then not very useful, unless the time corresponding to the minimum depth value has been quoted.

### 5.2 *1 D advanced dynamics models*

Usually dynamic models gives evaluation of velocity, assuming that the dynamic force on the vibrator is harmonic, corresponding to the frequency and the rotative mass moment.

At first, it must be quoted that the stability of the solution versus the phase of the dynamic force (velocity and displacement are supposed to be zero for initial step).

Secondly, experimental data show that the dynamic force include higher frequencies components while velocity is quite harmonic but with amplitude different of the 'free' value.

For 'a priori' prediction, analysis must include not only the so called dynamic and static mass of vibrator but also all the system above the sheet pile.

For experimental data analysis, the use of strain and velocity measurement at the top of the pile is fundamental. Analysis at different levels of the pile gives more accurate information. The analysis will be done on the so-called upward velocity and its evolution with the pile depth. Analysis of mean values is necessary both for strain and velocity. Mean velocity is easily derived from depth measurement Analysis of mean strain seems to be difficult.

Further works will propose comparison between experimental data and result of models. Direct analysis of strain and velocity will give evaluation of soil dynamic properties.

## REFERENCES

Ewing, W.M., Jardetzky W.S., Press F. 1957. Elastic Waves in Layered Media. McGraw-Hill Book Company.
ASTM Designation D4945-96. Standard Test Method for High–Strain Dynamic Testing of Piles.
Paquet, J. 1968. Etude vibratoire des pieux en béton, réponse harmonique et impulsionnelle. Application au contrôle. Annales de l'ITBTP, n°245.
Gonin, H. 1978. Etude théorique du battage des corps élastiques élancés et application pratique. Annales de l'ITBTP, n°361.
Gonin, H. 1996. Du pénétromètre dynamique au battage des pieux. Revue Française de Géotechnique n°76.

Application of Stresss-Wave Theory to Piles.,1988 Third International Conference. Ottawa, Canada, 25-27 May. Edited by Fellenius, B.H.

Screw piles. Installation and design in stiff clay. A. Holeyman, Editor.

Gonin, H. 1998. Quelques réflexions sur le vibrofonçage. Revue Française de Géotechnique n°83

Legrand, C., Van Rompaey, D. & Menten, J. 1994. A comparison of different sheet-pile installation methods. Proceedings of the 5th Intl. Conference on Piling and Deep Foundations, Brugge, pp.1.5.1-1.5.8.

Holeyman, A. & Legrand, C. 1994. Soil modelling for pile vibratory driving, U.S. FHWA International Conference on design and construction of deep foundations, Orlando, Florida, 6-8 december 1994.

Gonin, H. 1999. La formule des Hollandais ou le conformisme dans l'enseignement. Revue Française de Géotechnique n°87.

Jardine, R.J., Chow, F.C., Matsumoto, T. & Lehane B.M. 1998. A new design procedure for driven piles and its application to two japanese clays. Soils and Foundations, vol. 38 No.1,207-219, Japanese Geotechnical Society.

Holeyman, A. 2001. Engineering issues and modelling of vibratory driving. Assemblée Générale du Comité Français de Mécanique des Sols, 30 mai 2001, Paris, France.

Viking, K. 1997. Vibratory driven piles and sheet piles-a literature survey. Report 3035, Division of Soil and Rock Mechanics. Royal Institute of Technology. Stockholm.

Vipulanandan, C., Wong, D., Associate Members, ASCE, O'-Neill, M.W. 1990. Behavior of vibro-driven piles in sand. Journal of Geotechnical Engineering, vol. 116, No.8, August.

*Case histories and instrumented*
*vibratory driving*

# Sheet Pile Driving & Testing M5 East Project Sydney, Australia

S. Baycan
*Pile Test International, Australia*

ABSTRACT: A new 10km length of freeway in Sydney Australia, was opened in December 2001 with a total contract cost of $794 Million (AUD). As part of the project's ventilation tunnels, Vibro-pile (Aust.) Pty Ltd was contracted to drive sheet piles through challenging ground conditions. The sheets were designed to form part of the permanent cut & cover tunnel and were driven to depths of up to 20m. The sheets were required to resist both compression and tension loading. Early driveability predictions noted that the sheets would require impact driving to these depths after practical refusal to vibratory driving. In order to minimise project costs and meet time schedules, dynamic load testing of the sheet piles during impact driving was undertaken by Pile Test International (PTI). The construction method of using a combination of vibratory driving & impact driving and complementary dynamic testing was successful in completion of the contract to specification requirements.

## 1 BACKGROUND

The recently completed project allows efficient distribution of traffic through Sydney's southern suburbs. The project was a build, operate and maintain contract won by Baulderstone Hornibrook/Belfinger Berger (BHBB) Joint Venture in 1998, for the New South Wales Roads and Traffic Authority. As part of the ventilation system for Australia's longest road tunnel, the project involved the construction of sheet piles over a total plan length of approximately 300m, and sheets of up to 20m embedment. The tunnel was required to carry both compression loads from an overlying substation structure, as well as hydrostatic uplift pressures on the base slab. Vibropile (Aust.) Pty Ltd was commissioned by BHBB to install the sheets to nominated toe levels.

Geology of the area consists of deep alluvial variable (sands/clays) underlain by sandstone bedrock at variable depths. The project required testing of the sheet piles to prove their geotechnical resistance. An efficient method of "PDA" dynamic testing (Likins G, et al 2000) during impact driving of the sheets was proposed by Pile Test International and proved to be an effective means of achieving the required results.

## 2 SITE LOCATION AND GEOLOGY

### 2.1 *Site Location*

The site was adjacent to Wolli Creek in Arncliffe suburb of Sydney. The project was characterised by tight limits on construction noise and vibration due to the proximity of nearby residential buildings. Geology of the region consist of deep alluvial deposits, including layers of soft peat underlain by Sydney Sandstone bedrock.

### 2.2 *Geotechnical Conditions*

Geotechnical investigations by borehole drilling, standard penetration and cone testing by Golder Associates and Douglas Partners revealed variable consistency alluvial deposites with the top of the Sandstone bedrock at depths ranging from 10 to greater than 30m below ground level.

The groundwater level was reported to be approximately 1m below ground level.

Table 1. Soil Profile for Sheet Piles

| Soil Unit | Depth Range | Properties |
|---|---|---|
| Fill | 0-3m | Variable |
| Sand | 3-8m | Dense |
| Sandy Clay | 8-30m* | Residual |
| Sandstone | *Top rock varies 10 to 30m | Med strength |

The sands were generally loose, becoming dense, with the underlying clays generally sandy and of stiff to very stiff consistency. Layers of soft peat up to 1m in thickness were evident at below 10m depth.

The geotechnical report noted that the alluvial ground conditions closer to Wolli Creek, were more interbedded with sand and clay layers occurring regularly.

Cone penetrometer test results indicated the underlying clay materials recorded sleeve friction values of up to 300kPa.

## 3 SHEET PILE DESIGN & DRIVEABILITY

Based on the required permanent compression and tension loads on the sheet piles, they were installed to a nominated toe level. The sheets selected for the project were AZ18+1.0, 74.4kg/m weight, width=630mm and height=380mm and 390MPa grade.

Vibro-pile (Aust) Pty Ltd proposed the use of a vibratory hammer PTC, which is a Vibrofonceur 40A2(50Hz). This unit is powered by two number motors of 75 HP and operated at a constant 50 Hz frequency.

Early GRLWEAP driveability analyses performed by Douglas Partners (geotechincal consultants) indicated that sheets would not penetrate to the required depths by vibratory driving alone.

Preliminary analyses performed by PTI were undertaken to assess the likely level of effective refusal of the sheets to vibratory driving. Two types of analyses using the GRLWEAP program were carried out. At the time of the analyses, the type of hammer to be used was unknown. The simple models run assumed an ICE 815 hammer. The first of the analysis assumed the following;

- so called 'constant capacity option' with a resistance range of 400 to 1000kN.
- grab weight of 15kN and efficiency of 0.9.
- 70% shaft resistance.
- constant shaft resistance distribution top 8m, with a gradual increase between 8 and 10m, and constant below 10m depth.
- Shaft and toe damping values of 0.4 and 0.3s/m, respectively.

Figure 1 shows the results for the above analysis, which reveals that the pile is unable to penetrate or reaches effective refusal, (assumed for the purposes of this analysis to be at a penetration time of 900s/m), where a so called ultimate capacity of 700kN is reached.

To compare with the above, a 'penetrability analysis' using GRLWEAP with the following parameters was also undertaken;

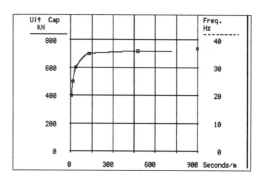

Figure 1. GRLWeap Analysis; Duoble AZ18+1.0 sheets using Vibratory Hammer

- Hammer details as per previous analysis.
- Shaft resistance top 10m (zero to 50kPa), 80 kPa between 10 to 15m, and >150kPa below 15m.
- Variation in shaft and toe damping parameters with depth, in accordance with the determined cone penetrometer soil profile.
- Relatively low toe resistance.

Results for the second analysis are summarised in Figure 2 and reveal the sheets to reach practical refusal (assumed at a penetration rate of 900s/m) at levels well above those of the final sheet toe level.

It is interesting to reflect on the early driveability analysis and use data gathered from the later impact dynamic tests carried out. In summary the following points are noted;

- the hammer model used was simplistic and most likely inaccurate to the one actually used on site.
- inadequate means of allowing for soil setup and clutch friction in the analyses.
- simplistic soil resistance distribution profiles, as the back-calculated damping factors and shaft and toe quakes from dynamic testing varied widely from those assumed in the analysis.

Despite the shortcomings however, preliminary driveability analyses confirmed intuitive estimates that the sheets were required to be impact driven to the required depths. Since an impact hammer was required, the decision was taken to drive piles in a "double pile" configuration i.e. a fully welded centre clutch.

For the double sheet pile sections used, Table 2 lists the design loads (i.e. the geotechnical ultimate);

Table 2. Soil Profile for Sheet Piles

| Location | Compression | Tension |
|---|---|---|
| Exhaust Duct | 500kN | 250kN |
| Fan Room | 540kN | 760kN |

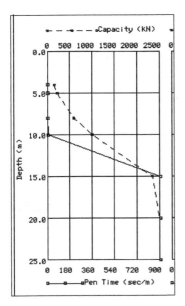

Figure 2. GRLWeap Analysis; Double AZ18+1.0 sheets using Vibratory Hammer

# 4 DYNAMIC LOAD TESTING DURING IMPACT DRIVING

## 4.1 Preliminary Testing

All sheets were initially advanced with vibratory driver up to 8m depth onto the top of the stiff, residual clay where the vibrator reached practical refusal. This was followed by impact driving to the top of the rock using firstly a 7 tonne for the shallow toe levels and in some cases a 9 tonne hydraulic hammer for the deeper sheet sections penetrating through the hard, residual clays.

Project specification required static load testing of the installed sheets. Due to time and cost constraints, PTI proposed PDA dynamic load testing of the sheets during impact driving to assess the load carrying capacity of the sheets.

Hartung, et al. (1992) used a similar method to test sheet piles and have reported extensive use of this method both within and outside Germany.

Hartung et. al (1992) reported some difficulties with lateral stress wave dispersion along adjacent sheets when the test section was within the actual wall structure. PTI specified that the test sheets be isolated from other sheets in the wall structure to avoid the lateral spread of the stress wave and to also avoid any positive influence of clutch friction on the resistance.

The measurement of bearing capacity of the sheets using dynamic testing was therefore a lower bound estimate of the actual in service conditions of the interlocked wall, which would include clutch friction effects.

Four sets of strain gauge and accelerometers were used to instrument the test sheets in the configuration as shown in Figure 3. The PDA instrumentation was attached to sheets near end of drive (EOD) and at restrike (RS). A large guide test frame was assembled to keep the sheets in alignment during high energy blows for testing.

Testing typically comprised taking 10 to 30 blows of data at high energy impact blows either at end of driven or restrike. Sample data shown in Figure 4, reveals that the sheet piles behave as a uniform elastic material. With known impedance, one dimensional wave mechanics implied in the PDA method is seen to apply. After successful early trial tests showing proportionality and sensible data acquistion using the PDA system, a program comprising a series of tests was compiled.

## 4.2 Test Program Results

A series of sheet pile tests were conducted in the period June to August 2000, to assess the available resistance both at end of drive (EOD) and at restrike. The test program and the results of Capwap analyses undertaken from the test blows is listed in Table 3.

The test sheets were restrike tested from 7 days up to 28 days post initial driving. Care was taken during high energy restrike blows to restrain the sheets in a frame structure so as to avoid buckling at the top and high out of plane stresses.

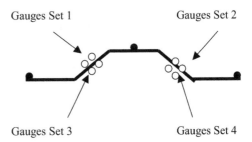

Figure 3. PDA Gauge Instrumentation on a Cross Section of a Typical Sheet Pile (schematic only-not to scale)

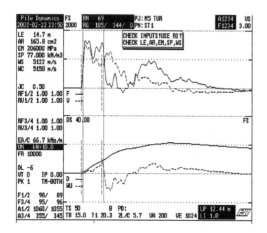

Figure 4. Sheet Pile Force/Velocity Trace from ST1 test

Table 3. Test Program Results & Capwap Resistance

| Sheet Pile | Final Penetration | Test Type | Friction (kN) | Base (kN) | Total Resistance (kN) |
|---|---|---|---|---|---|
| ST1 | 13.9m | EOD | 760 | 5 | 765 |
| ST2 | 12.0m | EOD | 910 | 1295 | 2205 |
| ST3 | 20.0m | EOD | 920 | 280 | 1200 |
| | | RS1 | 1880 | 295 | 2175 |
| | | RS2 | 2300 | 300 | 2600 |
| ST4 | 19.8m | RS1 | 2000 | 200 | 2200 |
| | | RS2 | 2300 | 300 | 2600 |
| ST5 | 14.3 | EOD | 610 | 305 | 915 |
| | | RS1 | 1080 | 220 | 1300 |

## 5 DISCUSSION OF DYNAMIC TEST RESULTS

Results showed that the ground conditions were highly variable with the portion of loads carried by the sheet pile base varying from only 1% to 60% of the total resistance.

Based on the (double) sheet base nominal cross sectional area, results for ST2 sheet which was driven to competent Sandstone rock, indicate a mobilised base resistance of 76MPa. It is interesting to note this high base resistance component, since this would generally be assumed to be minimal for sheet piles in normal design applications.

The distribution of friction resistance from the Capwap analyses concurred with the pile construction records in that the piles were vibrated up to 10m depth and then impact driven the rest of their embedment. A typical example of PDA data for a combination of friction and end bearing sheet pile ST2 is shown in Figure 5.

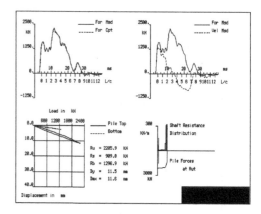

Figure 5. Capwap Analysis sheet pile ST2

Restrike testing on sheet piles revealed that up to 250 % increase in the available friction was achieved after less than 1 month of installation. Given the clayey nature of the ground conditions throughout the site, soil setup was expected.

The degree of soil setup however, provided additional assurance to the designers that the available friction would increase with time providing a higher factor of safety against any possible uplift forces.

Figure 6, shows the degree of soil setup with time. Although it was technically desirable to carry out a larger sample of restrike tests, due to construction time limitations, this was not possible.

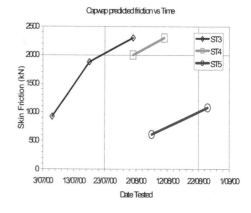

Figure 6. Restrike Test Results

## 6 CONCLUSION

The sheet pile driving project highlighted the current issues which related to sheet pile design, construction and testing. In particular, the author wishes to highlight the following points;

- Only a basic, evaluative type driveability analysis was undertaken to assess whether the sheets would penetrate the soils to the required depth with the vibratory driver.

- PDA testing proved to be an effective means of predicting capacity during the impact driving phase.

- The lack of a variable frequency vibratory hammer tended to limit the driveability of the sheets.

## 7 ACKNOWLEDGEMENTS

The author wishes to thank the direct and indirect assistance in compiling this paper to the following;

- Yahya Nazhat, Jeffrey & Katsuakas Consultants, formerly of Vibro-pile (Aust) Pty Ltd.
- Brian Ims, Douglas Partners Consultants.

## 8 REFERENCES

Hartung, M., Meier, K. & Rodatz, W. (1992) *Dynamic pile tests on sheet piles*, Proceedings of the Application of Stress-Wave Theory to Piles, F.B.J. Barends (ed.) Balkema, Rotterdam.

Likins.G, Rausche et al, PDA Pile Driving Analyser (2000) Manual, by Pile Dynamics Inc., USA

# Drivability prediction of vibrated steel piles

Noel Huybrechts & Christian Legrand,
*Belgian Building Research Institute (BBRI), Brussels, Belgium*

Alain Holeyman,
*Université Catholique de Louvain (UCL), Louvain-la-Neuve, Belgium*

ABSTRACT: Within the framework of a European research project (BBRI, 1994) and a subsequent national research project (BBRI, 1995-1997), the Belgian Building Research Institute investigated the vibratory driving technique for installing steel (sheet)piles. Drivability and environmental aspects of this installation method were studied, and two analytical models were developed in order to predict drivability and/or vibration nuisance in the surroundings. In this contribution the calculation method to predict the drivability of vibratory driven piles and sheet-piles is presented. The method is based on a simple analytical model integrating the equations of motion and calibrated by the results of a more complex wave equation simulation of the soil behaviour around the pile. The proposed method incorporates degradation of soils under cyclic loading as well as liquefaction of granular materials below the water table. Geotechnical input is derived from the results of cone penetration tests. The paper presents experimental results collected on several sites where geotechnical data and vibratory performance of sheet piles were collected. Correlation between calculated and measured driving times is ascertained.

## 1. INTRODUCTION

Three basic driving techniques can be used to install piles or sheet-piles : impact driving, vibratory driving and jacking. The method most commonly used in soft soils is vibratory driving because it usually allows a fast installation. Impact driving can produce high energy in difficult soil conditions, but is usually noisy and suffers from a low efficiency in common soil conditions; the energy is delivered by high forces induced by means of an impact hammer blow per blow. Jacking of piles can only be used where sensitive environmental conditions justify the rather high cost of this method; the driving potential comes from the reaction force of a dead weight or uplift resistance of already installed sheet-piles.

Driving is possible as long as the driving force, limited to a certain extent by available energy, exceeds the resistant forces (figure 1). Driving is possible as long as :

$$F_{driving} > F_{base} + F_{shaft} + F_{clutch} \qquad (1)$$

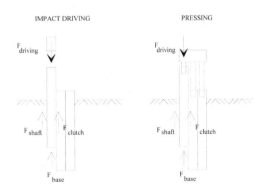

Figure 1. Drivability of impact driven and jacked (pressed) sheet-piles

Vibratory driving is often used to install steel (sheet) piles. Unlike impact driving and jacking where the resistance to be overcome is at least equal to the static soil resistance, installing (sheet) piles using vibration relies mainly on the reduction of the static soil resistance by vibrating the soil around the pile. As discussed by Dobry and Vucetic (1987) and Holeyman and Legrand (1994), the degradation of soil strength and the build-up of pore pressures as a result of cyclic movements lead to a significant reduction of the soil resistance from its static value. The ultimate build-up of pore pressure in saturated sands is referred to liquefaction, expressing thereby a complete loss of strength. Some consider that actually the pile is not installed into the ground primarily by the vibrating force but rather by sinking the pile into softened material under gravity forces.

## 2. THEORETICAL MODEL FOR THE DRIVABILITY OF VIBRATORY DRIVEN PILES

### 2.1. Model background

The drivability of vibratory driven piles has been widely investigated as part of a research program (BBRI, 1994). A detailed model was first developed based on the fundamental analysis of the dynamic behaviour of a cylinder embedded in a infinite medium (Holeyman, 1985). Cylindrical shear waves propagating away from the vibrated pile were evaluated using a one-dimensional radial discretization of the soil surrounding the pile. Elements of earthquake engineering normally used to asses liquefaction potential were applied to evaluate skin friction degradation upon cyclic shear stress. Degradation and excess pore-pressure were calculated based on correlation's derived from the friction ratio as measured in a CPT test (Holeyman, 1993b and Holeyman and Legrand, 1994).

A simpler analytical model has been developed to permit the prediction of the driving speed into the soil (Holeyman, 1993b). The simpler model discussed in the following sections allows one to quickly calculate the penetration speed at each depth of interest.

### 2.2. Estimate the acceleration amplitude

A first estimate of the acceleration amplitude (a) of the vibrating parts can be provided by the Eq. (2), assuming for now that the pile is a free body. Refinement of this first estimate will be performed later, taking into account soil reactions.

$$a = \frac{F_c}{M_v} = \frac{m_e \cdot \omega^2}{M_v} \tag{2}$$

where $F_c$ = centrifugal force of the vibrator
$M_v$ = mass of the vibrating parts (pile, clamps, vibrating part of the vibrator)
$m_e$ = eccentric moment of the vibrator
$\omega$ = frequency of the vibrator

### 2.3. Define the soil static resistance

The static base ($q_s$) and shaft ($\tau_s$) resistance profiles are derived from Cone Penetration (CPT) tests results, i.e. from the cone resistance $q_c$ and local unit skin friction $f_s$ (E1 cone).

### 2.4. Calculate the soil liquefied resistance

The totally liquefied base ($q_l$) and shaft ($\tau_l$)unit soil resistance are derived from the prior step based on an exponential law as expressed respectively by equations (3) and (4):

$$q_l = q_s \cdot [ (1-1/L) \cdot e^{-1/FR} + 1/L ] \tag{3}$$

$$\tau_l = \tau_s \cdot [ (1-1/L) \cdot e^{-1/FR} + 1/L ] \tag{4}$$

where $q_l$ = liquefied soil base resistance
$\tau_l$ = liquefied soil shaft resistance
FR = friction ratio as measured in a CPT test with E1 cone (% of the mantle friction to the cone resistance, i.e. FR = 100 $f_s$ / $q_c$)
L = empirical liquefaction factor expressing the loss of resistance attributable to liquefaction (L will be higher for saturated and loose sands and is chosen in the range of 4 to 10)

90

## 2.5. Calculate the soil driving resistance

The driving base ($q_d$) and shaft ($\tau_d$)unit resistance are derived from the static and the "liquefied" soil resistance depending on the vibration amplitude following an exponential law as expressed respectively in equations (5) and (6).

$$q_d = (q_s - q_l) \cdot e^{-\alpha} + q_l \qquad (5)$$

$$\tau_d = (\tau_s - \tau_l) \cdot e^{-\alpha} + \tau_l \qquad (6)$$

where $q_d$ is the driving base unit resistance, $\tau_d$ is the driving shaft unit resistance and $\alpha$ is the acceleration ratio (a/g) of the pile, as obtained from Eq. (2)

At each depth z the vibratory pile driving resistance is calculated :

$$F_{base} = q_d \cdot \Omega \qquad (7)$$

$$F_{shaft} = \chi \cdot \int_{z=0}^{z=D} \tau_d \cdot dz \qquad (8)$$

where $\Omega$ is the pile section, $\chi$ the pile perimeter and D the pile penetration.

## 2.6. Recalculate the vibration amplitude

The soil resistance causes a reduction of the vibration amplitude initially assessed for a free body. Equation (2) can be modified into :

$$a = \frac{m_e \cdot \omega^2 - \delta \cdot F_{shaft}}{M_v} \qquad (9)$$

where $\delta$ is a damping factor expressing the ability of the shaft resistance to dampen the movement of the sheet-pile

The above mentioned calculation steps (Eq. (3) through (9)) can be cycled through until the difference between two consecutive values of the vibration amplitude becomes negligible.

## 2.7. Calculate the driving speed

The vibratory driving speed is obtained by applying Newton's law on the vibratory pile driving process (fig. 2).

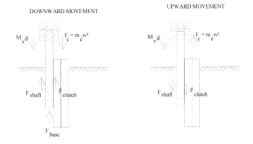

Figure 2. Drivability of vibratory driven sheet-piles

The net effective downward and upward forces are calculated as :

$$F\!\downarrow(t) = m_e \cdot \omega^2 Cos(\omega t) + M_v g - F_{base} - F_{shaft} - F_{clutch} \geq 0 \qquad (10)$$

$$F\!\uparrow(t) = m_e \cdot \omega^2 Cos(\omega t) - M_v g - F_{shaft} - F_{clutch} \geq 0 \qquad (11)$$

Finally, the resultant driving speed is calculated by integrating the net effective downward and upward accelerations over a cycle of duration $T = 2\pi / \omega$ :

$$v = v\!\downarrow + v\!\uparrow = \int_0^T F\!\downarrow(t) / M_v \cdot dt - \int_0^T F\!\uparrow(t) / M_v \cdot dt \qquad (12)$$

## 3. COMPARISON OF PREDICTED AND OBSERVED DRIVING TIMES

A computer program has been developed based on the general approach explained above. Some simplifications, which are not addressed in this paper, have been introduced to facilitate and speed up the evaluation of equation (12). Although the model presented herein is rather simple, a very good agreement is obtained between the predicted and the recorded driving times on a variety of job sites with different soil characteristics and different vibratory hammers and piles. The predicted and the observed penetration times are compared in Table 1.

From the last two columns in Table 1, it can be seen that the measured penetration times correspond well to the predicted in a lot of cases (by comparing predicted and measured values, one has to take into account that site heterogeneities can lead to differences in the measured values up to 50% and more), especially in those cases were, due to the subsoil conditions, the vibrating technique is very efficient , e.g. in loose to medium dense fine sands and weak cohesive soils.

An example of such a site is the Hingene test site in Belgium, at which an extensive measurement campaign was conducted. Figure 3 shows the subsoil profile as provided by a CPT test (cone E1) at the test site. The water table was encountered at 3.1 m depth. The test results showed good agreement between predicted and observed penetration times and between the predicted and measured vibration levels at both ends of the sheet pile during installation. At this site, comparisons between predicted and observed installation time with different sheet pile lengths and different hammers confirm the validity of the suggested model.

In other cases though, the predicted penetration times deviate from the observed ones. An example of such a site is the Kortrijk Site in Belgium, where a 20.6 m long tubular steel pile with a thickness of 9.5 mm and a diameter of 1 m had to be installed. Figure 4 shows the subsoil profile as depicted by a CPT test (cone M4) performed at the site. The water table was encountered at a depth of -1.8 m. The upper twelve meters consist of very soft river deposits; below a sandy layer, was found a very stiff tertiary clay layer into which the tube had to be driven.

A preliminary calculation with the above described computer program pointed out that the necessary time to install the pile to a depth of 20 m with a PTC 30HFV vibratory hammer was 14½ minutes or could lead to refusal at a depth of 18.2 meters (depending on the assessment of the local friction). However, driving met refusal at a depth of about 11 m.

The reason for the difficult driving and the difference between the predicted and the observed penetration speed was explained by measurements taken during the actual driving of the pile. The pile vibration amplitude was measured by means of a velocity transducer placed at the pile head and a velocity transducer (protected by a cover) at the pile toe. Figure 5 shows the monitored amplitude of vibration at the pile top and at the pile base upon loss of drivability. The observed frequency was 38 Hz.

From the measurement results, one can observe that :
- the vibration amplitude at the pile top (0.65 mm zero to peak) is considerably less than the nominal vibration amplitude which is,

$$\frac{m_e}{M_{vibr}} = \frac{m_e}{M_{hammer} + M_{pile}} = \frac{26000 kg.mm}{(6500 + 4820) kg} = 2.3mm$$

- the amplitude at the pile base (0.45 mm zero to peak) is smaller than the amplitude at the pile top (0.65 mm)

It would appear that the pile base amplitude (0.45 mm) is not sufficient to allow the pile to penetrate as the stress-strain behaviour for clayey soils is primarily elastic for small amplitudes. The small vibration amplitude may be related to one or more of the following considerations :
- An important soil (i.e. clay) mass is sticking to the vibrating pile, leading to a more important vibrating mass, leading to a smaller vibration amplitude.
- The vibratory hammer may not be able to deliver the required energy, and thus not maintain its nominal amplitude or frequency. A characteristic of the variable eccentric hammers is that a lack of power results in a reduction of vibration amplitude (rather than a reduction of frequency (Houzé, 1994)).
- A smaller amplitude at the pile base is obtained due to the elasticity of the pile.

Table 1. Predicted and observed penetration times

| Site | Subsoil | Vibrator type (f [Hz]/m_c [kg.m]) | Pile or sheet pile Type | Length [m] | Install. depth [m] | Predicted penetration time [min' sec"] | Measured penetration time [min' sec"] |
|---|---|---|---|---|---|---|---|
| Hingene (B) | Fine loose to compacted sand/clay | PTC 50H2 (25/50) | AZ 13 | 9,8 | 9 | 1' 44" | 1' 55" |
| | | | BZ 17 | 11,7 | 10 | 2' 31" | 2' |
| | | | BZ 17 | 16,1 | 14 | 5' 35" | 3' 40" à 4' |
| | | PTC 30HF (36/26) | BZ 17 | 11,7 | 10 | 2' 03" | 2' 20" |
| | | | BZ 17 | 16,1 | 14 | 4' 31" | 6' 35" |
| | | PTC 30HFV (38/26) | BZ 17 | 16,1 | 14 | 4' 15" | 6' à 8' |
| St-Stevens- Woluwe (B) | silt or clay | PTC 50H2 (25/50) | AZ 26 | 14,5 | 13 | 3' à 4' (*) | 3' 30" |
| Watermaal- Bosvoorde (B) | loose sand | PTC 13HF (34/13) | BZ 12 double | 8 | 6,4 | 55" | 32" |
| | | | BZ 12 simple | 8 | 6 | 29" | 11" |
| Moerzeke (B) | Weak clay/compacted sand | PTC 15HF (38/15) | AZ 18 | 11,5 | +/-11 | 2' 30" à 3' 30" (*) | 15' à 16' until 10.6 m (refusal on the compacted sand layer) |
| North-Sea | clay | ICE 1412 (22.5/115) | Open tube (Ø 1.067 m) | 28 | | Refusal at 8.5 m | Refusal between 10 & 12.75 m |
| | | | Open tube (Ø 1.067 m) | 27 | | Refusal at 9 m | Refusal between 10 & 12.75 m |
| Kortrijk (B) | clay | PTC 30HFV (38/26) | Open tube (Ø 1.01 m) | 20,6 | 20 | 14' 36 until 20 m à refusal at 18.2 m after 18' (*) | Refusal at 10.7 m after 25' (free) |
| | | PTC 110 HD (20/105) | Open tube (Ø 1.01 m) | 20,6 | 11 tot 20 | 3' à 5' (*) | 5' 25" until 19.25 m +/- 8' to 9' until 20 m (free) |
| Boortmeerbeek (B) | sand/clay | PTC 30HFV (38/26) | Larssen IIIN | 12 | 11,4 | 2' 28" until 11.4m à refusal at 10.3 m after 8' (*) | 10' à 15'. half of it for the last 3 m (clay) |
| Koksijde (B) | sand | PTC 15HFV (38/15) | Omega 8 | 7 | 6,5 | 30" | 46" |
| | | | Omega 8 | 8 | 7,5 | 35" | 55" |
| Wielsbeke (B) | loamy sand | PTC 23HFV (38/23) | AZ 18 | 12,5 | 10,5 | 1' 04" à 1' 23" (*) | +/-2' |
| Gent-Zeehaven (B) | sand-clay/loam | PTC 25H2 (28/25) | AZ 13 | 18,5 | 17,5 | 5' 44" à 9' 20" (*) | 5' à 6' 30" |
| Genk (B) | sand | DIESEKO 50M (28/50) | Larssen 2S | 14 | 12,6 | 5' 12" à 7' 51" until 7.8 m (*) | 12' until 7.8m |
| Vilvoorde (B) | loam-sand gravel-clay | DIESEKO 50M (28/50) | Larssen 7 | 14 | 16 | 3' 12 à 3' 50" (*) | 6' 20" |
| Gent-Ringvaart (B) | sand-loam/clay | ICE 23RF (38/23) | ZN41/1670 | 14,5 | 15,5 | 1'47" à 1' 58" until 12.5m (*) | +/- 1' until 12.5 m |

Table 1 (continued). Predicted and observed penetration times

| | | | | | | | |
|---|---|---|---|---|---|---|---|
| Anderlecht (B) | loam/clay | ABI VRZ 600GL (37/8.6) | DWZ 135 | 10 | 9 | 2' 46" à refusal at 8.6 m in 5' 38" (*) | 1' 28" (free) 3' 21" (with clutch friction) |
| Landen (B) | loam | ABI VRZ 600GL (37/8.6) | AZ13 | 7 | 6,7 | 30' (38 Hz) à 1'04" (33 Hz) | 1'18" |
| Erembodegem (B) | loam/sand | PTC 30HF3 (32/26) | BZ17 | 10,7 | 10 | 1' 56" à 2' 25" (*) | 3' 15" (free) 3' 40" (with clutch friction) |
| Hamburg (D)(**) | sand/clay | Müller MS-25 HF (26/25) | closed tube Ø 273 mm ST2 | 17 | 17 | Refusal at $q_c$ peaks (sand) at 2m and 9m | 4' |
| | | Müller MS-25 HF (28/25) | closed tube Ø 356 mm ST3 | 12 | | refusal at 1 m | Refusal at 5m |
| | | Müller MS-25 HF (28/25) | Open tube Ø 356 mm ST3 | 12 | 5 tot 12 | 54" (with friction 0-5 m) 1'15" (without friction 0-5 m) | 1' |
| | | Müller MS-25 HF (28/25) | Open tube Ø 356 mm ST4 | 17 | 17 | 4' 24" | 1' |
| Luik (HsA) (B) | gravel/schist | ICE 14 RF (35/14) | BZ17 | 9,8 | 7,5 | 4' 18" | 3' 10" |
| | | ICE 14 RF (37/14) | BZ17 | 12 | 8,2 | 8' 10" | 4' 40" |
| | | ICE 14 RF (39/14) | PU25 | 11,6 | 8,2 | 2' 07" | Refusal at 8 m after 7' 40" |
| | | ICE 416 (17/23) | BZ17 | 12 | 8,2 | Refusal at 6.9 m (5m reached after 8') | Refusal at 5 m after 6' |
| Limelette (B) | loam | ICE 14 RF (38/14) | PU16 (S) | 14 | 13 | Vibratory driving possible until 13 m (8m reached after 2' 16") | Refusal at 8 m after 10' |
| Kortrijk-Leie (B) | alluvial clay | Dieseko PVE2315 (35/15) | PU16 (D) | 12 | 11,5 | 4' à 11' (*) | 3' à 3' 20" (free) |
| Etten-Leur (N) | compacted sand/ clay | Dieseko VM40 (33/40) | AZ18 | 19 | 18,5 | Refusal at 9.9 m, which is reached after 7' | +/- 10' until a depth of 9 m, after which penetration becomes very difficult (up to 1 hour driving before installation depth is reached) |
| | | ABI MRZV40 (43/8) | PU32 (D) | 14,2 | 13,5 | Refusal at 9.1 m after 9' | Refusal at 9 m, which is reached after 6' à 8' |
| | | | PU16 (D) | 13 | 12,5 | 'Refusal at 9.3 m after 2' à 3' | Meas. 1 : 4' until 8m, after which refusal Meas. 2 : 3'45" until 8m and 6'50"until 10 m, after which refusal |

(*)   If the CPT is performed with mechnical cone, then calculations are made by deducing the local friction from the total side friction measurement and from the cone resistance
(**)  After Westerberg et al. (1995)

94

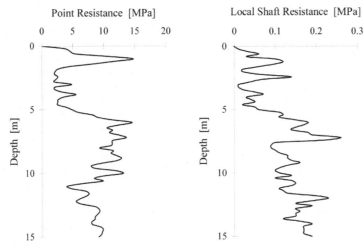

Figure 3. Subsoil profile at Hingene site (B)

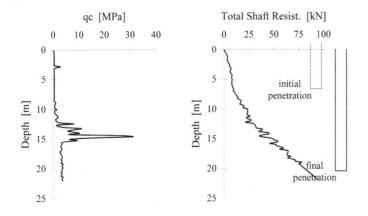

Figure 4. Subsoil profile site at Kortrijk (B)

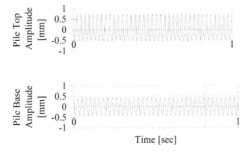

Figure 5. Record from the vibration amplitude upon refusal

By applying the observed vibration amplitude to the calculation model (Fig. 6), a much better correlation between the calculated and the observed penetration time was obtained. The pile was placed at the bottom of an excavation at -2,5 m and penetrated another 4,5m under its own weight. As a result, observed and calculated penetration rates are reported starting at level - 7 m. Figure 6 evidences that the difference for the predicted and observed penetration times for the site in Kortrijk was not due to an incorrect estimation of the dynamic soil resistance but due to an incorrect estimation of the vibration amplitude, which happened to be limited by the nominal power of the generator.

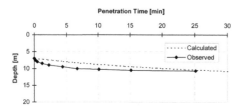

Figure 6. Predicted and observed penetration times site at Kortrijk

## 4. EVALUATION AND LIMITATIONS OF THE SUGGESTED MODEL

Although the model presented herein is rather simple, the calculated driving times show good agreement with the observed driving times under a variety of subsurface conditions. We may conclude that the calculation of the dynamic soil resistance may well show very good agreement with physical reality. In some cases however, e.g. in stiff clay layers or very dense sand layers, predictions deviate from the observations.

Analysing in more detail refusal conditions reveals that the determination of the dynamic soil resistance is not the only part governing the drivability calculation.

Based on experience with different sites (BBRI, 1994 & 1995-1997), possible refusal conditions attributable to the vibratory hammer, to the piles, and to the soil are summarised hereunder. Not all these phenomena are incorporated in the calculation model, thereby evidencing the limitations of the suggested approach.

*Attributable to the vibratory hammer :*
- The cyclic movement reduces the static soil resistance allowing the pile to penetrate under its steady driving forces. Lack of penetration speed can be due to the insufficient weight of the non-vibrating part of the vibratory hammer.
- An insufficient eccentric moment of the vibratory hammer leads to an deficient vibration amplitude.
- Deficient power of the driving powerpack leads to a decrease of the vibration amplitude or of the vibration frequency of the vibratory hammer.
- Deficient serviced hammers will not be able to run at their nominal amplitude and frequency.

*Attributable to the pile :*
- Too heavy a pile reduces the vibration amplitude following the equations (1) and (8). On the other hand , a heavy pile is advantageous as appears from equation (10) and (11). A compromise has to be found.
- When the pile is too elastic (e.g. too small pile section $\Omega$), the vibration amplitude can be insufficient at the pile base, too much transverse vibration can occur preventing the pile from penetrating, and the pile can be damaged.
- Amongst other possible problems, poor quality sheet-piles lead to an important increase in clutch friction resistance.

*Attributable to the soil :*
- Too large a base resistance (e.g. rock) leads to bouncing of the vibrated pile.
- The soil resistance can increase due to compaction (e.g. sand) as reported more in detail by Youd (1992).
- The dynamic soil resistance (e.g. hard clay) may not be sufficiently degraded. Some references deal with this subject (Dobry and Vucetic, 1987; Holeyman and Legrand, 1994).
- Soil sticking to the vibrating pile (e.g. stiff clay) can reduce the pile vibration amplitude.
- Too elastic a soil stress-strain behaviour leads to a lack of relative displacement of the pile with regards to the soil (e.g. clay).

The summarized observations do not cover all possible drivability problems : in many cases, phenomena different from those monitored are interfering, as illustrated by the analysis of the Kortrijk results.

## 5. CONCLUSIONS

A calculation method has been developed for the prediction of vibratory driven piles and sheet-piles. The model allows to calculate the penetration time and to define whether it will be possible or not to drive a pile or sheet-pile to the required depth. The best choice regarding drivability between the available hammers can be made by calculating the driving times for different hammer types in given site conditions. The comparison of the predicted penetration times with the observed times demonstrates the reliability of the model.

It has been shown that a correct calculation of the dynamic soil resistance during the vibratory driving process is one of the important issues in the determination of the drivability. Further research is required to analyse phenomena leading to refusal.

## REFERENCES

Dobry, R. & Vucetic, (1987). M. State-of-the-art report : Dynamic properties and response of soft clay deposits, *Proceedings of the International Symposium on Wave Propagation and Dynamic Properties of Earth Materials*, Vol. 2, pp 51-87, Mexico City, 1987.

Holeyman, A. (1985) "Dynamic non-linear skin friction of piles," *Proceedings of the International Symposium on Penetrability and Drivability of Piles,* San Francisco, 10 August 1985, Vol. 1, pp. 173-176.

Holeyman, A. (1993a) "HIPERVIB1, An analytical model-based computer program to evaluate the penetration speed of vibratory driven sheet Piles", *Research report prepared for BBRI,* June, 23p.

Holeyman, A. (1993b) "HIPERVIBIIa, A detailed numerical model proposed for Future Computer Implementation to evaluate the penetration speed of vibratory driven sheet Piles", *Research report prepared for BBRI*, September, 54p.

Holeyman, A. & Legrand, C. (1994). Soil Modelling for pile vibratory driving, *International Conference on Design and Construction of Deep Foundations*, Vol. 2, pp 1165-1178, Orlando, U.S.A., 1994.

Houzé, C. (1994). HFV Amplitude control vibratory hammers : piling efficiency without the vibration inconvenience, in DFI 94, pp. 2.4.1 to 2.4.10, *Proceedings of the Fifth International Conference and Exhibition on Piling and Deep Foundations*, Bruges, Belgium, 1994.

BBRI .(1994). HIgh PERformance VIBratory pile drivers base on novel electromagnetic actuation systems and improved understanding of soil dynamics, *Progress reports of the BRITE/EURAM research contract CT91-0561*, 1994.

BBRI (1995-1997). Heibaarheid van damwanden en diepfunderingen d.m.v. intrillen – ontwerpaspecten en omgevingsaspecten, IWONL-conventie CC CI 253B/6148/CC CIB 253B/6049, IWT-conventie RD/95/05, DGTRE-conventie C.3214-01.

Van Rompaey, D., Legrand, C., Holeyman, A., 1995. A prediction method for the installation of vibratory driven piles, *Soil dynamics and earthquake engineering conference*, Crete, Greece, may 1995.

Westerberg, E, Eriksson, K., Massarsch, K.R. 1995. Soil resistance during vibratory pile driving, *Proceedings of the International Symposium on Cone Penetration Testing*, Volume 3, Linköping.

Youd, L.T. (1972). Compaction of sands by repeated shear straining, *Journal of the Soil Mechanics and Foundations Division*, 1972, Vol. 98, SM7, pp. 709-725.

# Vibrodrivability and induced ground vibrations of vibratory installed sheet piles

Viking K.
*Royal Institute of Technology (KTH)*

ABSTRACT: The vibratory characteristics of full-scale sheet piles, vibratory driven into the ground, have so far not been described systematically in detail. This paper concerns the preliminary results of a series of full-scale field tests where both the driveability and induced ground vibrations have been continuously monitored. Furthermore, the results of measured ground vibration levels have been measured at two different sites characterised as having hard and easy driving conditions. In the light of the presented results, the paper briefly discusses how the complexity of a driveability and vibration prediction can be subdivided into three parts, namely, vibrator-, sheet pile-, and soil-related parameters, all of which significantly affect both the driveability and induced ground vibrations.

## 1 INTRODUCTION

Vibratory-driving techniques are used all over the world primarily for driving and extracting sheet piling. Even though the technique is global, there are very few publications relating specifically to using vibratory driving, compared to the number relating to using impact driving, for piles or sheet piles. The lack of publications in the area reveals both the lack of knowledge and the potential for further research.

A major limitation with using the vibratory technique for driving piles is a lack of guidelines for vibratory driving in relation to driving refusal and bearing capacity. This has lead to the current situation where vibratory-driving techniques are used primarily for driving and extracting sheet piles.

From a conducted literature survey, Viking (1987), it was found that the vibratory characterisation of full-scale sheet piles, has not been published systematically in detail.

Presented paper is a brief summary of Viking (2002), containing some of the obtained results in relation to a limited full-scale field tests conducted in the end of 1999.

The full-scale tests were conducted as part of a research project at the Division of Soil- and Rock Mechanics, Royal Institute of Technology (KTH).

## 2 PRIMARY MECHANISMS OF VIBRO-DRIVEN SHEET PILES

The following paragraph is divided into these particular sections based on the conclusions of the literature survey (Viking 1997), from which it can be concluded that today's research into the use of vibratory techniques can be described as a number of isolated islands of knowledge. This paragraph is therefore an attempt by the author to briefly present these islands of knowledge using a common language, as a hypothesised overview of the vibro-driveability of sheet piles.

In light of the complexity associated with attempting to describe vibro-driveability, it is justifiable to subdivide the prime factors affecting vibro-driveability into the following four subsections: (i.) kinematics of the system, (ii.) vibro-related factors, (iii.) sheet pile related factors, and (iv.) soil-related factors.

### 2.1 *Kinematics of the system; vibrator, sheet pile, and soil*

The ease with which a sheet pile enters a soil stratum depends on the characteristics of the mechanical interaction and dynamic nature of the whole vibrator, sheet-pile profile and soil system.

The kinematic nature of the whole vibrator, sheet pile and soil system described here is based on the following assumptions and simplification: (i.) the bias weight is assumed not to vibrate, (ii.) the sheet pile is simplified to behave as a rigid body (i.e. the head and toe have the same acceleration), (iii.) the sheet pile and excitor-block of the vibrator are also simplified into a rigid body having the same acceleration as the sheet-pile head, and (iv.) there are no soil and sheet-pile phase differences.

As a consequence of the above assumptions and simplification, the kinematics of the whole system can be simplified and "idealised" according to Figure 1. The dynamic forces being referred to, act on the "idealised" vibrator, sheet pile and soil system mentioned above as illustrated in Figure 1 can be obtained by establishing a dynamic equilibrium during the downward and upward stroke of the penetration motion.

## 2.2 Vibro-equipment related factors

The most important factors related to the vibrator-equipment as part of the system are generally the choice of vibro-parameters such as driving frequency etc., (not discussed here). However, this section only briefly discusses the role of the efficiency of the equipment.

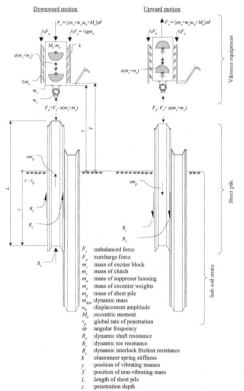

Figure 1. The major components of the vibrator, sheet pile and soil system, illustrating the free-body diagram of the direction of the forces acting on the system during downward and upward stroke motion.

### 2.2.1.1 Efficiency ratio of actual to theoretical driving force

From the conducted literature study (Viking 1997) it appears as though the efficiency of the vibrator-equipment significantly affects the optimum vibro-driveability. Moulai-Khatir et al. (1994) have stated that in field conditions, the efficiency can be estimated to be in the range $0.20 < x < 0.25$.

The efficiency factor ($\xi = F'd / Fd(t)$) in this study has been defined as the ratio between the amplitude of the actual force delivered (F'd) over the theoretical driving force (Fd(t)). Where the theoretical force is calculated according to the vibrator specifications and operating range which is expressed by:

$$F_d(t) = F_0 + \left[M_e + (m_v + m_c)u_0\right]\omega^2 \cos(\omega t) \qquad (1)$$

were the magnitude of the actual delivered (read measured) driving force is taken from the readings of the strain gauges mounted near the sheet-pile head.

At this point, it must be noted that the actual peak force delivered to the sheet-pile head will always be less than the theoretical peak driving force, due to the simple fact that non-negligible energy losses exist in each part of the vibrator equipment, and also due to the flexural and torsional energy propagation in the relatively slender sheet-pile profile, which is not sensed by transducers that measure only axially-applied loads.

## 2.3 Sheet pile related factors

The aim with the following subsection is to position the author's overall view on the key factors related to the behaviour of key factors related to the sheet pile that affects the vibro-driveability.

### 2.3.1.1 The axial sheet pile rigidity assumption

The primary objectives of this section of the paper is to discuss why and when it can be appropriate to treat and model a full-scale pile or sheet pile as a rigid body during vibratory driving. In other words, defining when there will be no significant longitudinal vibrations having any engineering impact on the vibro-driveability.

Viking and Bodare (1998) used the following rule of thumb when defining a system with a vibrator and model pile of length L behaving as a rigid body:

$$\frac{1}{4 f_d} = \frac{T}{4} \geq 2 . t_n \qquad (2)$$

where $T = fd-1$ the time period of the unbalanced force Fv, and tn = 2L/cb the time taken for a stress wave to travel back and forth.

Which can be explained phenomeno-logically by the following. If it can be assumed that a vibratory-driven sheet pile tends to behave like a rigid body when the time it takes for the driving force to change from zero to maximum (T/4) is greater than or equal to twice the time (tn = 1/fn) it takes for a tentative stress wave to travel the distance (4L) along the sheet pile, having a bar velocity of (cb), (see Figure 2).

Analysis of the forced longitudinal vibrations of vibratory-driven sheet piles can be considered to closely follow the conditions of a free and longitudinally-vibrating finite rod of length L, constrained at the longitudinal position (x = 0) by the vibrating force produced by the vibrator, and free at its other end (x = L), (see Figure 2). The resonance frequency (fn) of a vibratory-driven sheet pile, can there for be estimated by the time (tn) it takes a stress wave to travel the distance (2L) of the sheet pile with a bar velocity of (cb), (see Figure 2).

In order to illustrate this, assume that a vibratory-driven sheet pile is 14 m in length, with a (cb) for steel of ~5100 m/s. The natural resonance frequency (fn) of the vibrating sheet-pile can be estimated by Equation (2) where the first mode (n = 1) gives rise to a resonance frequency of f1 ~ 182 Hz.

It is therefore justifiable from an engineering point of view, to see a vibratory-driven sheet pile as behaving like a rigid body. Since longitudinal resonance phenomena seem to generally occur at very high frequencies compared with the driving frequencies (fd) used by today's modern vibratory-equipment (usually in the range of 30 < fd < 40 Hz). Such justifiable assumption would greatly simplify the problems associated with a single-degree of freedom vibrating system.

Effects of induced lateral vibrations on vibro-driveability

Transversal and flexural mechanisms are generally ignored in both vibro-driveability and environmental analysis, which usually confines itself to just longitudinal behaviour. However, these lateral effects may at times have such a major engineering impact that they should be considered during pre-analysis of both vibro-driveability as well as the ground vibrations induced. It is also fairly unexpected that in the most often cited article in the context of induced ground vibrations (Clough el al. (1980)), that the existence of lateral effects is not even mentioned to exist.

The primary objectives with this section is to discuss why and when it is appropriate to consider the effects of dynamic lateral motions of vibratory-driven slender profiles. These laterally-induced motions generally arises due to following situations, or combinations of these; (i.) U-profiles driven one at a time by vibrators equipped with a single clamping device holding the profile at the web, (ii.) when the vertical alignment is neglected when threading the new interlock to the pre-installed profiles leading interlock, (iii.) choice of vibrator equipment and the way the equipment is operated, and (iv.) when the presences impurities and permanent deformations of the interlock occurs.

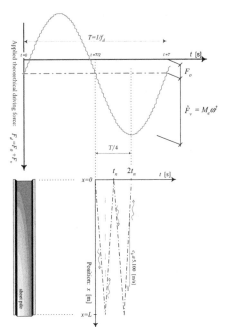

Figure 2. Illustration of the relationship for rigid-body behaviour expressed by Equation 2.

The shape of the lateral curvature of the driven sheet pile is very sensitive to the end restraints, which is nicely illustrated when "idealising" the effects of lateral motion to the case when comparing a leader-mounted vibrator with a free-hanging vibrator system (see Figure 3).

2.3.1.2 Effects of friction force in sheet-pile interlock
It should be noted that it is not currently possible to theoretically quantify the interlock friction force (Rc) developed during installation, due to the complexity of the problem describing when, how, and to what extent the way the vibro-equipment is operated together with how the presence of soil grains in the interlock affects vibro-driveability and the ground vibrations induced. It is also challenging attempting to describe the whereabouts and the extent of the effects of obstacles such as boulders that may be present in the subsoil strata. These obstacles can cause damage to the leading interlock, drift from initial verticality, and sometimes clutch release. All of these are basic factors known to have significant impacts on the dynamic interlock friction force (Rc).

From field-test-related results in relation to these issues, Viking and Bodare (2000), on the induced ground vibrations, it was found that ground vibrations can be two to five times the magnitude of equivalent situations without (Rc). It should be noted that the conducted studies were of a purely research nature, featuring brand new sheet piles, and didn't consider any production-capacity-related aspects, which concludes that the presence of (Rc) has in fact

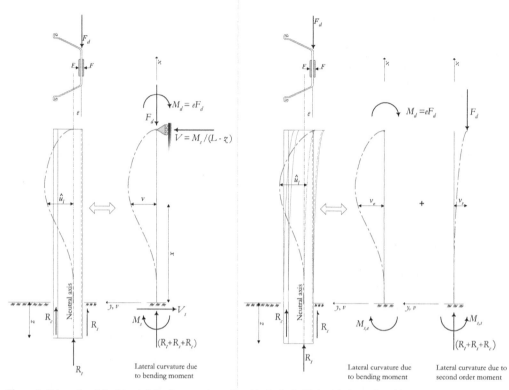

Figure 3. Schematic of the lateral deflection curvature of an ideal, elastic U-shaped sheet pile with a leader-mounted vibratory-system to the left and a free hanging system to the right.

a no negligible impact on the ground vibrations generated.

### 2.4 Sub-soil related factors

The aim in the following subsections is to position the author's overall view on the key factors and fundamental mechanisms of the soil behaviour in the vicinity of a vibratory-driven sheet pile. It should be noted that the following subsections do not represent the final solution as to how the intrinsic soil behaviour should be addressed. This is due to the complexity of the problem, and the fact that the soil behaviour as it is hypothesised later on, has not been scientifically verified due to a lack of appropriate soil instrumentation during the field tests.

2.4.1.1 Fundamentals behind the shear strength reduction

The shear strength reduction of cohesionless soils can be idealise to studies of the behaviour of the interaction of individual granular particles in a regular array in order to point out the key phenomena. The stresses developed within the idealised granular soil volume, either by external loads at the boundary or by the weight of the soil, generate intergranular nor-

mal (N) and tangential (T) contact forces between the individual soil grains, as illustrated by Figure 4a.

Several authors have implied that the shear-strength reduction induced is primarily related to the soil mechanism termed liquefaction. However, the shear strength reduction is also present under artificial conditions such as in air-dried sand in laboratory tests performed by Barkan (1962), Bernhard (1967), O'Neill and Vipulanandan (1989a), and Viking (1998). From this it can be concluded that soil liquefaction is not the primary parameter for the induced shear-strength reduction in a vibratory-technique context.

As the vibratory-driven sheet pile undergoes a longitudinal acceleration amplitude (a) in the range of 10-20g, it interacts with the soil volume in the vicinity of the profile. The interaction between the sheet pile and the soil introduces inertial forces to the soil. The induced inertial forces causes a dynamic motion of the individual grains since the inertial force exceeds the magnitude of the initial vertical confining stress (sv) within the soil. As a consequence, the shear stress level decreases as the individual grains become separated from one another, the grains within the array starts to experience "free fall" and the short-time drops in the intergranular contact forces N and T. The almost total absence of vertical confining stress (sv) within the soil volume during parts of each steady-state loading appears to be the key phenomenon behind the shear-strength reduction during vibratory-installation of sheet piles, as illustrated in Figure 4b.

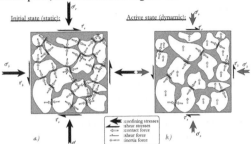

Figure 4. A cubically-packed assemblage of soil grains illustrating how effective confining and shear stresses are carried by the static inter-granular contact forces.

Even though it appears as though the cyclic acceleration induced motion of the individual soil grains is the key mechanism behind the shear strength reduction, it is of course evident that induced excess pore-pressure is undoubtedly of great assistance in reducing the shear strength during the installation process in field conditions. It seems reasonable to correlate the large amplitude of cyclic accelerations (cyclic strains) that generate large volume changes (De) of a cyclic nature, which in turn produce a cyclic variation in the induced excess pore-pressure (Du). As the cyclic acceleration amplitude (cyclic strain) in the soil is assumed to decrease with increasing radial distance (r), it is most likely that the amplitude of the cyclic variation of pore-pressure changes (Du) will decrease in the same manner, with increasing radial distance.

## 3 FIELD TESTS

The full-scale field tests reported on here were performed in Vårby, a suburb of Stockholm (Sweden). The test site was chosen firstly for its relatively homogeneous soil conditions, and secondly because there was good probability of being able to keep the sensors in place for the entire duration of the planned field tests.

The two main objectives of the measurements were; (i.) to study and document the actual magnitude of field-related parameters such as driving force, dynamic soil resistance, and ground vibrations generated with and without the presence of clutch friction, with both the axial and lateral accelerations of the sheet pile, and (ii.) to find out whether it was possible to correlate a prediction of sheet-pile driveability with two existing driveability models found during the literature survey (not presented here).

### 3.1 Soil conditions

Apart from the top 1.5-2.0 m layer of topsoil and clay, the soil conditions at the Vårby test site consisted of the more than 40 m of medium-dense to almost loose glacial sand. The ground water table lies approximately 2.0 m below ground level, and the 40 m of glacial sand is relatively well-graded, varying between a silty and a gravelly sand, with depths shown in Table (4-2).

The soil investigations performed at the test site Vårby included; (i.) three CPTu type tests, (ii.) 21 plus an additional 14 dynamic probing tests, and (iii.) soil sampling at six different levels.

Furthermore, pore pressures were measured using a piezometer and an open stand-pipe. Results of the dynamic-probing tests were obtained using the recommended Swedish standard equipment for dynamic probing, and further details can be found in Axelsson (2000).

The results of the three CPTu tests, revealed a sandy deposit that is relatively homogeneous with respect to the range of areas investigated and penetration resistances, see Figure 5.

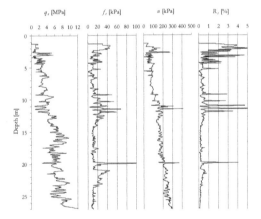

Figure 5. Results of CPT tests from the Vårby test site.

The soil profile at the test site was interpreted from the CPT tests and the soil sampling at six different levels, further details can be found in Axelsson (2000), here presented in Table 1.

Table 1 Soil profile (after Axelsson, 2000).

| Depth range [m] | | | Soil type |
|---|---|---|---|
| 0 | → | 2.5 | Clay |
| 2.5 | → | 4.5 | Silty sand |
| 4.5 | → | 8 | Sand |
| 8 | → | 14 | Silty sand |
| 14 | → | 19 | Gravelly sand |
| 19 | → | 24 | Sitlty sand |
| deeper than 24m | | | Sand |

### 3.2 *Vibrator equipment*

The vibratory-driver system used to install the sheet piles during the full-scale field test at Vårby was a leader-mast-mounted ABI vibrator of type static moment variable model (MRZV 800V), mounted on an ABI telescopic leader mast (TM 14/17 L), which were held by a remodelled Sennebogen excavator.

Information of basic performance data of used vibrator equipment can be obtained through the web page of the vibro-manufacturer, (www.abi-gmbh.de).

### 3.3 *The sheet piles*

There were two main criteria for selecting the sheet pile profiles for the field tests. Both the length and the profile should be representative of the most common vibratory-driven sheet piles in Sweden. The typical sheet piles were determined to be steel sheet-piles from Profil ARBED S.A. (Luxembourg), 14 m in length.

The following is a summary of the number and type of sheet piles used; (i.) three sheet piles of profile LX-16, 14 m in length, without sensors, and (ii.) one sheet pile of profile PU-16, 14 m in length, fitted

with sensors, driven several times, with the positioning of the sensors shown in Figure X.

Information of dimensions and other relevant data about the two types of sheet-pile profiles used are provided by the sheet pile be obtained through the two sheet pile manufacturer, (Profil ARBED and British Steel).

### 3.4 *Instrumentation*

The instrumentation system used to document vibro-driveability and the environmental effects generated during the tests consisted of the following five main parts; (i.) vibrator instrumentation, (ii.) the sheet-pile sensors, (iii.) soil instrumentation, (iv.) the data-acquisition system, and (v.) the video camera.

The instrumentation system together with the dynamic data-acquisition system are illustrated in Figure 6. The instrumentation setup consisted of 25 channels in total, subdivided into the following three groups; (i.) the vibrator, (ii.) the sheet pile, and (iii.) the tri-axial geophones.

The vibrator-related instrumentation monitored the following three parts: (i.) the static surcharge

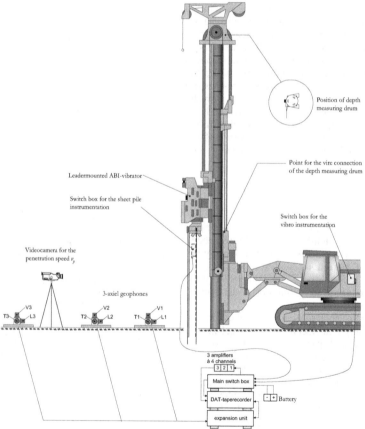

Figure 6. Illustration of the instrumentation and data-acquisition system.

force (Fo) applied (the oil-pressure), (ii.) the penetrative movement of the sheet pile (z) (the penetration depth), and (iii.) the adjusted static moment (Me) (i.e. the position of the eccentric weights).

The sheet-pile-related instrumentation consisted of 10 strain gauges and three accelerometers, positioned as illustrated in Figure 7. Eight of the 10 strain gauges were mounted in holes cut into the sheet-pile section and later filled with a two-component resin in order to protect the sensors. The accelerometers were small enough to fit into the holes. One of the three accelerometers was mounted laterally in order to monitor the laterally-induced motions in the sheet pile.

The soil instrumentation consisted of three tri-axial geophones, see Figure 6. The purpose of these was to monitor the ground vibrations induced during the installation phase.

The data-acquisition system was developed especially for these tests and consisted of a switch box, 3 four-channel strain-gauge amplifiers, and one DAT tape recorder (SONY‰ PC216Ax) together with a channel-extension unit (SONY PCCX32Ax), see Figure 6. The dynamic data recorded from the various sensors were analysed as discussed in Section 4.

## 3.5 *Sheet pile sensors*

The arrangement of the sheet-pile instru-mentation was critical to the success of the full-scale field tests. Redundancy was purposely factored into the design of sheet-pile sensor system, due to the inherent risk of losing sensors or cables during the installation phase.

At the sheet-pile toe (known to be the most critical position), four strain-gauge circuits all wired as full Whetstone bridges, were mounted in holes cut out for them (see Figures 7 and 8). The purpose of the four circuits was to monitor the magnitude of the dynamic soil resistance (Rt) generated at the sheet-pile toe. The holes were cut out with precision as close to the toe as possible, but without excessive risk of losing them if boulders were encountered during the test. The holes provided at the same time protection to the sensors and at the same time they did not change the pile-surface geometry, since the holes later on were filled with a two-component resin in order to protect the sensors.

At the sheet-pile head, two strain-gauge circuits were mounted on the inside and outside surfaces of the sheet-pile web. The purpose of these two circuits was to monitor the driving force (Fd) generated by the vibratory equipment and at the same time possible to document effects of bending moments induced in the sheet pile (as illustrated in Figure 3).

The three low-g accelerometers, from Analog Devices Inc., were mounted on specially developed printed circuit boards, small enough to fit into the

cut out holes of the sheet pile section. Two of the tree accelerometers were mounted in the longitudinal direction, one at the head and at the toe, see Figure 7. The third accelerometer was laterally mounted at the mid position of the 14 m-long sheet pile The purpose was to be able to monitor the laterally induced motions of the sheet pile, shown in Figure 6, which significantly affected both driveability and induced ground vibrations.

### 3.5.1.1 . The tri-axial geophones

The soil instrumentation consisted of three tri-axial geophones in order to monitor the induced ground vibrations during the installation phase. The tri-axial geophones were mounted on a steel plate equipped with soil-spikes with the purpose of transferring the induced ground vibrations to the geophones. The relative position and radial distances from the sheet piles being driven are illustrated by Figure 6.

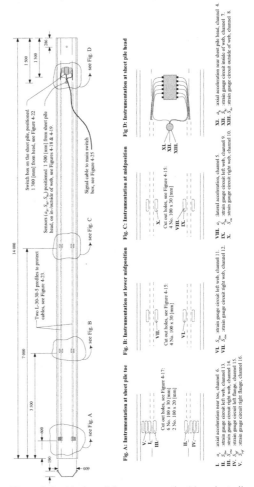

Figure 7. Positioning of the sensors on the 14 m sheet-pile profile PU-

105

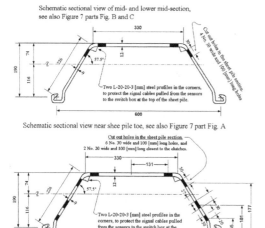

Figure 8. Positioning of the strain gauges used in the sheet-pile sections.

3.5.1.2 Visual monitoring of displacement amplitudes

Two kinds of stickers were stuck onto the sheet piles at every meter in order to visually monitor both the vertical and lateral double-displacement amplitude. Visual readings were taken each metre as the stickers and sheet pile entered the soil strata. The axial and lateral double-displacement readings together with the penetration depth (z) m were orally noted on a micro cassette recorder.

The principle used here, is that the human eye is able to make visual readings when the velocity is zero, that is at the end points of the sinusoidal motion. The visual reading generates a fuzzy picture of two stickers instead of one. Readings of the double-displacement amplitude of 12 mm are taken at the cross-section of the two fuzzy stickers (horizontal scale and the slope) generated as shown on Figure 9.

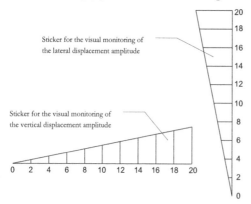

Figure 9. Stickers used for visually monitoring the sheet-pile displacement amplitude, below the visualisation of the ocular reading of the 12 mm vertical displacement amplitude.

# 4 RESULTS

Observation of the time histories of the sheet pile head and toe forces together with induced environmental effects of the full-scale field tests have provided a unique opportunity to develop a better understanding of the penetrative mechanisms and the generated ground vibrations.

## 4.1 *Vibro-driveability*

Figure 10 displays the results of vibro-driveability in two different ways, namely the t-z curve (on the left-hand side), and the vp-z curves (on the right-hand side of the each field test), from which the following observations can be made.

Overall, the penetration speed (vp) is relatively constant during the whole installation phase (130-140 mm/s), which should be seen in light of the fact that all vibrator parameters were kept constant, and that the soil conditions at the Vårby site were relatively homogeneous. However, a distinct dip in the penetration speed (vp) (read as lower values on vp) during the initial depth range of approximately 1-3 m were present during both situations, i.e. with and without effects of friction forces in the interlock. It is speculated that the initial dip in vp could be explained by the presence of an initial lateral flexibility of the sheet pile, resulting in a lower vibro-driveability (read as lower values of vp), since the dip is so pronounced during the initial 2 m. This speculation is partly supported by the recordings of significantly higher peak values by the laterally-mounted accelerometer. Another possible explanation for this is that the dip could be related to the initial approximately 2 m thick topsoil layer.

## 4.2 *Forces and accelerations*

The two curves denoted by Shi and Sho in Figures 11a and b showing generally higher peak force values, represent the driving force amplitude (Fd) actually transmitted to the sheet-pile head. The four curves denoted by Srtf , Sltf , Srtw and Sltw in the same two figures, generally display a lower peak force amplitude, and represent the peak value of the dynamic soil resistance measured at the toe (Rt).

The two Figures 12a and b displays the recorded results of the three recorded peak accelerations (ah , at and al) values with and without the effects of friction forces in the sheet pile interlock plotted versus penetration depth.

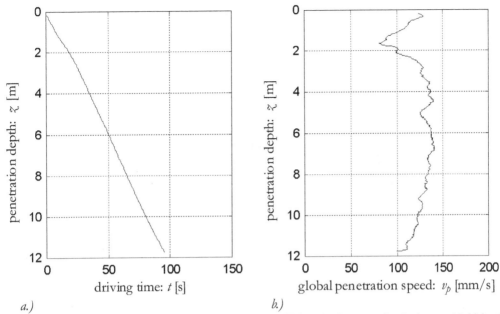

a.)                                          b.)

Figure 10. Vibro-driveability of sheet pile without presence of clutch friction, showing penetration depth versus (a) driving time, and (b) global penetration speed.

The two curves denoted by (ah) and (at) in Figures 12a and b, show a generally higher acceleration amplitude, and represent the axial accelerations of the sheet-pile head and toe respectively. The curve denoted al represents the lateral acceleration of the vibratory-driven sheet pile. The peak acceleration versus depth curves with friction forces (Rc) present in the sheet-pile interlock, displays a more jagged pattern of the acceleration curves as a reflection of the presence of friction forces in the sheet-pile interlock. The acceleration curves not only displayed a more jagged pattern, but they also displayed higher peak acceleration values. The peak axial accelerations in the case of freely-driven sheet piles (without the effects of friction force (Rc)) were in the vicinity of 15 g for the head acceleration (ah), and 17 g for the toe acceleration (at) during the entire depth range investigated (see Figure 12a).

The difference observed between peak head and toe accelerations (ah and at) during the most favourable conditions was found to be in the range of 2-3 g. The minor differences between the peak head and toe acceleration recordings (ah and at) graphed in Figure 12a, ought to be viewed as a confirmation of the assumption of pile rigidity, as previously discussed. The slightly higher values of the toe acceleration (at) compared with the head acceleration, are most likely explained by inertial effects on the 0.6 m long sheet-pile mass below the position of the toe accelerometer (see Figure 6).

As previously discussed, the evaluated ratio of actual to theoretical driving force (x = F'd /Fd(t)). The theoretically delivered peak compressive force (Fd(t)) is evaluated to $(70/2 + 6(82p)2 + 15g2.450 \sim 794$ kN, according to Equation 1. The surcharge force (Fo) used in these calculations is 35 kN, which is half the capacity of the push-down force of the equipment used, and the deceleration of the sheet-pile head was measured to ~15 g (see Figure 12a). The peak unbalanced force (Fv) is evaluated to 361 kN, with a static moment set to half of the maximum specified performance data of the vibrator.

The peak compressive force measurement (F'd) is approximately 310 kN and evaluated from Figure 12a. The ratio x is then evaluated as $(310/794 \sim 0.39)$, which exhibit values agreeing with Moulai-Khatir et al. (1994).

From which it can be concluded that the measured value (read as the value actually delivered) of the peak force at the sheet-pile head was clearly less than the theoretically computed value. This fact is most likely explained by energy losses in the vibrator-sheet-pile system, and most likely to have been affected by the induced lateral motion of the vibratory-driven sheet pile.

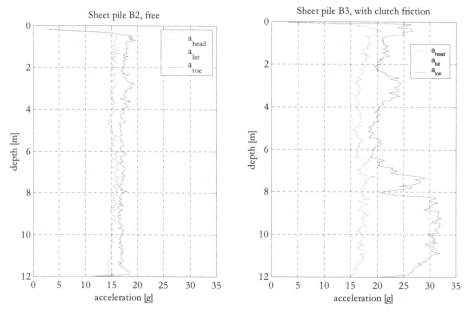

Figure 11. Dynamic peak amplitude of section forces at the head and toe versus penetration depth; where (a) is a test without Rc , and (b) is a test with Rc present.

### 4.3 Ground vibrations

The three graphs denoted (a) on the left-hand side of Figure 13, show the monitored peak vibrations without the effects of clutch friction, and the corresponding graph denoted (b) show the corresponding results including the effects of clutch friction forces in the sheet-pile interlock, from which the following primary observations can be made.

From the general analysis of all the test results, it was concluded that the presence of clutch friction produces peak values were approximately twice as great as those without the effects of friction forces.

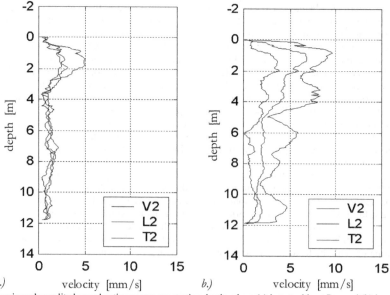

Figure 12. Dynamic peak amplitude acceleration versus penetration depth; where (a) is test without Rc , and (b) is a test with effects of Rc present.

The previously mentioned decrease in penetration speed (vp) during the initial depth range (1 < z < 3 m), corresponds to an equivalent observation of a significant increase in monitored peak particle velocities versus depth. Why these effects occur and how these observations should be interpreted is not immediately obvious at present.

From the quite extreme value of the vertical peak particle-vibrations monitored approximately 50 mm/s (not presented here), it can be inferred that the individual soil grains in the vicinity of the measurement positions is exhibiting a vertical acceleration of approximately (50 * 10-3 * (2p41)) which is approximately 1.3 g. The implication of this in relation to internal shear strength reduction of the soil, is somewhat scientifically vague due to the inadequate soil-instrumentation. However, could be considered as an indication that the relationship between the effects of inertial forces on the individual soil particles and the internal shear strength reduction of cohesionless soils (as previously discussed), is somewhat true.

### 4.4 *Dynamic load transfer curves*

Further insight into the penetration mechanisms of sheet piles during vibratory driving can be obtained by observing the load transfer relations (Rt - u and Rs - u curves). Figures 14 and 15 are taken from the evaluated time histories of Test without the presence of clutch friction, simply because the phase lag between head and toe acceleration, and the section forces at both the head and the toe were found to be close to zero, and the time window chosen corre-

sponds to a penetration depth of approximately 3.1 m.

The curves (al , ah , at and Sltf ) in Figures 14a-c appears to be in phase and the peak value of the lateral displacement can be evaluated to about (ûl = âl/w2 ~ 50/(2p41)2 ~ 1 mm), which at a first glance does not appear to be of significance, however is about 40 per cent of the evaluated peak axial acceleration.

It can be seen from the time 'history' of the dynamic section force (Sltw) representing the dynamic toe resistance (Rt) in Figure 14f, that the curve is biased around a negative value instead of varying in a pattern between approximately zero and a positive peak value, corresponding to a compressive loading during the downwardly- directed part of the penetration motion. Initially it was believed that the negatively biased values were attributable to inertial forces acting on the 600 mm long part of the sheet pile below the position of the transducers mounted near the pile toe. However, another and more likely explanation for this biased value is the fact that from an engineers point of view, it is fairly obvious that the bending and twisting of the sheet pile is the likely explanation for the results obtained.

The few earlier published reports and plottings of the relationship of dynamic load transfer curves of the toe resistance, (Rt-z), relationship that do exist include O'Neill and Vipulanandan (1989), Dierssen (1994), and De Qock (1998); all of which relate to slow vibratory-driving of displacement piles.

The general pattern of the few slow Rt-z curves reported, is a concave, upward loading-unloading pattern, indicating the development of strain harden-

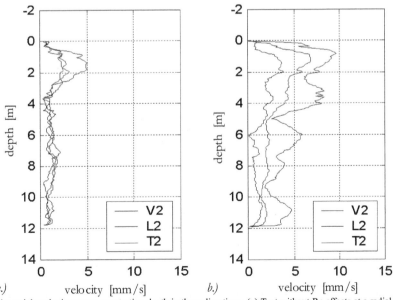

Figure 13. Peak particle velocity versus penetration depth in three directions: (a) Test without Rc effects at a radial distance (r2) of 4.1 m and (b) Test with Rc effects at an (r2) of 4.15 m.

109

ing. Another observation from the few slow Rt-z curves published, is that they never tend to reach the typical plateau seen in classic elastoplastic patterns (similar to the Smith soil model of impact loading of the toe).

It has by the author been hypothesised (Viking 2002) that slow Rt-z curves of vibratory-driven, low-displacement piles (sheet piles and H-beams) ought to display an Rt-z curve similar to displacement piles, but with a less developed, concave-shaped, upward loading and unloading curve. This is explained by having a different penetration mechanism (low-displacement pile penetration) which is probably more related to plunging than displacement. The pattern of Figure 15b gives a kind of indication of this being the case, however it's quite obvious that further studies needs to be conducted before anything can be concluded.

## 5 CONCLUSIONS

The process of obtaining representative values for the dynamic soil properties is the most difficult part of describing and develop a better understanding of

the vibro-driveability and the clutch friction developing in the sheet-pile interlock.

It should once more be emphasised that there are surprisingly few publications available that present full-scale testing of vibratory-driven sheet piles, and is especially surprising considering the numerous kilometres of sheet piles driven each year. It can therefore be concluded that today's engineering knowledge about the vibratory-driving technique is still in its infancy. Therefore there is a considerable need to perform more studies to allow definite conclusions to be drawn and adequate theories to be developed.

The difficulties encountered in attempting to shed some light on the difficulties of describing the vibro-driveability, have justified the simplification to divide the prime factors affecting vibro-driveability into three main categories: (i.) vibratory-equipment-related, (ii.) sheet-pile related, and (iii.) soil-related factors.

The following factors have been concluded to affect the vibro-driveability; (i.) Efficiency of the vibratory-equipment: any successful attempt to simulate vibro-driveability should begin with a reasonably accurate theoretically-generated value for

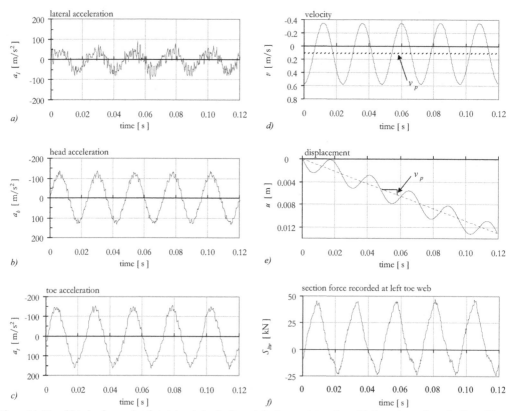

Figure 14. Time histories for acceleration, integrated velocity and displacement, together with dynamic toe force for Test without clutch friction effects, at a penetration depth (z) of ~ 3.1 m.

110

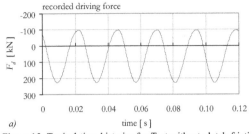
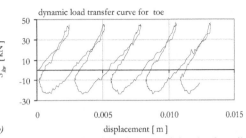

Figure 15. Typical time histories for Test without clutch friction effects, where (a) is the driving force recorded at the sheet-pile head, and (b) the evaluated dynamic load transfer curves at the sheet-pile toe.

the driving force that actually enters the sheet-pile head axially, (ii.) Justification of axial rigidity: from an engineering point of view this can be justified in most of the more favourable cases featuring favourable soil conditions, (iii.) Effects of lateral flexibility in the sheet pile: this factor can definitely cause unexpected problems, such as considerably lower penetration speeds and the generation of considerably higher ground-vibration values, (iv.) Effects of friction forces in the clutch: these could also cause considerably lower production capacity, generate considerably higher ground-vibration values, and in the long run, generate damaging settlements, (v.) Inertial forces but not primarily the effects of liquefaction: these have been phenomenologically explained as the hypothetical fundamentals of shear-strength reduction. They are the primary soil-related factors that alter the initial shear strength of the soil to the favourably-reduced value during the process of vibratory-installation of sheet piles.

## 6  COMMENTS

It should once more be emphasised that there are surprisingly few publications available that present full-scale testing of vibratory-driven sheet piles, and is especially surprising considering the numerous kilometres of sheet piles driven each year. It can therefore be concluded that today's engineering knowledge about the vibratory-driving technique is still in its infancy. Therefore there is a considerable need to perform more studies to allow definite conclusions to be drawn and adequate theories to be developed.

The process of obtaining representative values for the dynamic soil properties is the most difficult part of a tentative attempt to predict vibro-driveability. It is the author's opinion that a new field investigation method and analysing procedure is needed to enhance the process of obtaining more representative values for the dynamic soil resistance in the future.

From a geotechnical point of view, subsoil characteristics are usually determined by means of stan

dard investigation methods such as; (i.) probing tests (for example CPT and SPT), (ii.) sampling tests (for example boring), and (iii.) laboratory tests (such as tri-axial, resonant-column, and direct shear tests).

Generally speaking, all these methods have been developed to produce input information for static design issues. It is obvious that such investigation methods are not well suited to characterising the soil behaviour that exists during the vibratory-installation process of piles and sheet piles. However, it is obvious that the investigation method devised to evaluate the dynamic soil stress and strain characteristics that exist in cohesionless soils must attempt to duplicate the boundary conditions existing in the vicinity of vibratory-driven sheet piles as closely as possible.

## 7  ACKNOWLEDGEMENTS

The author would like to express his sincere gratitude to The Development Fund of the Swedish Construction Industry (SBUF) together with the Swedish construction company Skanska AB for funding the research project.

The full-scale field tests could not have been accomplished without the advice, and supply of materials from the following individual and firms; Håkan Bredenberg, Kent Allard at Geometrik AB, Thomas Merz at ABI Gmbh, ProfilARBED, Skanska Grundläggning AB, Carl-Oscar Nilsson and Jonas Green for their invaluable assistance with the performance of the experiments, Christian Legrand and Noel Heubrecht at BBRI, and R.J. van Foeken, to all I'm very grateful.

## 8  REFERENCES

Axelsson, Gary., (2000), Long-Term Set-up of driven Piles in Sand., PhD thesis 1035, Div. of Soil and Rock Mechanics, Institute of Technology, Stockholm, Sweden, 194 pp.

Clough, G.W., Chameau, J-L., (1980), Measured Effects of Vibratory Sheetpile Driving., Journal of the Geotechnical Engineering Division, ASCE, Vol. 104, GT 10 pp. 1081-1099.

De Cock, Sébastian., (1998), Comparison de modèles de vibrofonçage avec des résultats de labora-toire et de chantier., Travail de fin d'études présenté pour l'obtention du dimplôme d'ingénieur Civil des Constructions, Dept. AMCO, Univ. Catholique de Louvain, Belgium, pp. 113.

Dierssens, G., (1994), Ein bodenmechanisches Modell zur Beschreibung des Vibrations-rammens in körningen Böden., Doctoral Thesis, University of Karlsruhe, Germany.

Moulai-Khatir, Reda., O'Neill, Michael W., Vipulanandan, C., (1994), Program VPDA Wave Equation Analysis for Vibratory Driving of Piles., Report to The U.A. Army Corps of Engineers Waterways Experiments Station., Dept of Civil and Environmental Engineering, UHCE 94-1, Univ. of Houston, Texas, August 1994, 187 pp.

O'Neill Michael W., Vipulanandan, C., (1989), Laboratory evaluation of piles installed with vibratory drivers., National Cooperative Highway Research Program, Report No. 316, National Research Council, Washington, DC. Vol. 1. pp. 1-51. ISBN 0-309-04613-0.

Viking, K., Bodare, A., (1998), Laboratory studies of dynamic shaft resistance response of a vibro-driven model pile in granular soil by varying the relative density., Proc. to XII European Conference on Soil and Foundation Engineering, 6 pp.

Viking, K., Bodare, A., (2000) Proc. to Deep Foundations Institute 25th Annual Meeting Conf. and 8th International Conf. on A Global Perspective on Urban Deep Foundations., New York, New York USA, October 5-7, 2000, 19 pp.

Viking, K., (1997), Vibratory driven piles and sheet piles -a literature survey., Report 3035, Div. of Soil and Rock Mechanics, Institute of Technology, Stockholm, Sweden, 75 pp, ISSN 1400-1306.

Viking, K., (2002), Vibro-driveability -a field study of vibratory driven sheet piles in non-cohesive soils., PhD thesis 1002, Div. of Soil and Rock Mechanics, Royal Institute of Technology, Stockholm, Sweden, 281 pp, ISSN 1650-9501.

*Deep soil vibrocompaction*

# On-line compaction control in deep vibratory compaction

Wolfgang Fellin & G. Hochenwarter
*Institute of Geotechnical and Tunnel Engineering, University of Innsbruck, Austria*

A. Geiß
*Bauer Spezialtiefbau GmbH, Schrobenhausen, Germany*

ABSTRACT: Deep vibratory compaction (vibroflotation) is a method of ground improvement up to depths of 40 m. It has been successfully used since 1936. The major problem of this method is, that the process of compaction is not well understood, although it works very well. The methods of compaction control during the vibration are unreliably. The achieved compaction can only be verified after vibration.
In this paper the development of the new monitoring and quality control system "Bauer-RDV-online" is presented. It is shown that information from the movement of the vibrator can be used as additional quality control and indicator for the degree of compaction. Such information can be analyzed directly during compaction and offers thereby the possibility of an "on-line compaction control". This is shown based on theoretical studies and recent results of successful full-scale tests in Germany.

## 1 VIBRATORY COMPACTION

Vibratory compaction is used for underground improvement of non-cohesive soils, namely sand and gravel with less than 25% fines and less than 5% clay. The grains are brought into a closer packing. Thus the density of the soil is increased, as well as the friction angle and the stiffness. The bearing capacity is increased, settlements are reduced and the permeability of the soil is decreased. In addition the behavior of the soil becomes spacial more uniform. Compaction of loose sandy soils below ground water level reduces the risk of liquefaction in the case of an earthquake.

The vibrator is a cylindrical tube that houses eccentrically rotating masses. These rotating masses induce a horizontal vibratory motion (Fig. 1). The tube diameter is between 25 and 42 cm, its length between 2.7 and 4.4 m, the weight between 800 and 2600 kg. The induced centrifugal force is from 150 up to 472 kN with a frequency from 25 up to 60 Hz.

The vibrator is lowered due to its vibration and its own weight. Water or air jetting can increase the penetration rate. Typical penetration rates are 1.0 up to 2.0 m/min. When the required depth is reached, the water flow is reduced. There are two methods of withdrawing the vibrator (Fig. 2):

• Withdrawing in steps.

• Withdrawing and lowering in cycles (pilgrim

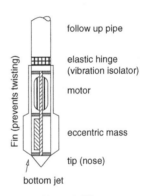

Figure 1. Vibrator.

step method).

Backfill material is added into the developing crater. Up to 1.5 m³/m depth might be required. Sand and gravel are proper backfill materials, it is chosen similar to the in situ soil.

The usual maximal depth is about 25 m but sometimes it is possible to reach 40 m. The upper 3 m are poorly compacted. They have to be removed or compacted using other compaction methods, e.g. heavy tamping or roller compaction.

The compacted soil column of a single vibration

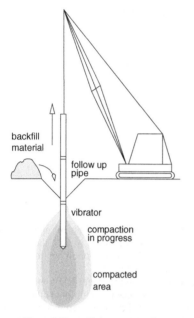

Figure 2. Deep vibratory compaction.

point has a radius of 1.5 to 3 m. This radius depends on the soil and the vibrator. It can only be estimated by experience, there is no theoretical approach up to now. To compact an area the vibrator is lowered at points in a rectangular or triangular pattern, for spread footings and compaction of large areas, respectively. The spacing between the compaction points is usually between 1.5 and 3.0 m. Patterns for spread footing have shorter spacings than patterns for large areas. The finer the sand the smaller is the spacing.

For further information see e.g. (D'Appolonia 1953; Poteur 1968; Brown 1977; FG Straßenwesen 1979; Broms 1991; Fellin 2000c; Smoltcyk 2001).

## 2 COMPACTION CONTROL - STATE OF THE ART

Verifying the compaction success is quite difficult because the density of cohesion less soil below the surface cannot be measured directly. Other quantities like cone penetration resistance have to be used to determine the achieved density. Compaction control is done in two steps.

### 2.1 Monitoring during vibration

During compaction the following parameters are recorded over the depth:

- Quality and amount of backfill material

- Maximum and time history of power consump-

tion

- Time needed for each compaction point

- Consumed energy for each compaction point

The power consumption is usually used to deduce the achieved density. This is by no means sufficient!

### 2.2 Compaction control after vibration

The achieved density is tested by cone penetration or standard penetration tests in several centers of the triangles or rectangles of the compaction pattern. It should be noted that the penetration resistance measured immediately after compaction can be up to 50% lower than the penetration resistance measured one month later. This is the case mainly in fine sands below ground water level.

## 3 ON-LINE COMPACTION CONTROL

Vibrocompaction is influenced by two major components. The *vibrator* on one hand and the surrounding *soil* on the other hand. If the characteristics (e.g. the density) of the surrounding soil change during vibration, also the motion of the vibrator must change. This is the fundamental idea of the on-line compaction control. The motion of the vibrator can be analyzed based on simple physical and soil-dynamic models (Fellin 2000a; Fellin 2000b; Fellin 2000c).

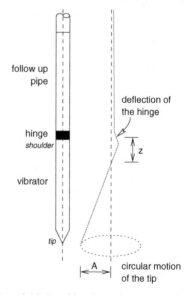

Figure 3. Motion of the vibrator freely hung up in air.

116

## 3.1 Vibrator in air

The motion of the vibrator freely hung up in air can be easily observed. This can be used to verify and calibrate the measurement equipment. Observations show the rotary motion represented in Figure 3, with a deflection of the tip and the shoulder. The simplest model of a vibrator oscillating in air is the physical pendulum sketched in Figure 4. The plane motion of this pendulum is a projection of the rotation of the vibrator.

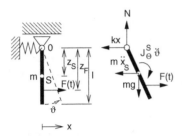

Figure 4. Physical model of the vibrator in air.

For an excitation force $F(t) = F \cos \Omega t$ and a vanishing spring stiffness of the support, as well as under the assumption that the natural frequency of usual vibrators is much smaller than their operating frequency, it applies to the amplitude of the tip:

$$A_L = \frac{F}{\Omega^2} \left( \frac{1}{m} + \frac{(z_F - z_S)(l - z_S)}{J_\Theta^S} \right) \tag{1}$$

The total mass $m$ is the sum of the not rotating masses $m_v$ of the vibrator and the eccentric mass $m_e$, which rotates in the distance $r$. The moment of inertia $J_\Theta^S$ around the center of gravity $S$ has to be calculated concerning both masses. The induced centrifugal force is $F = m_e r \Omega^2$.

## 3.2 Vibrator embedded in soil

The interaction between vibrators and soil can be modeled in analogy to a foundation oscillation (Fig. 5).

Two decoupled oscillations constitute the motion, namely a horizontal oscillation $x_S$ and a rocking oscillation $\vartheta$. The natural frequencies are $\omega_x = \sqrt{k/m}$ and $\omega_\vartheta = \sqrt{k_\vartheta/J_\Theta^S}$:

$$x_S = X_S \sin(\Omega t + \alpha_S)$$

$$X_S = \frac{F}{\sqrt{(k - m\Omega^2)^2 + c^2 \Omega^2}} \tag{2}$$

$$\alpha_S = \arctan\left( \frac{k - m\Omega^2}{c\Omega} \right)$$

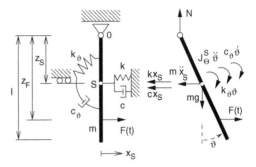

Figure 5. Analogy to foundation oscillation.

$$\vartheta = \Theta \sin(\Omega t + \alpha_\vartheta)$$

$$\Theta = \frac{(z_F - z_S)F}{\sqrt{(k_\vartheta + mgz_S - J_\Theta^S \Omega^2)^2 + c_\vartheta^2 \Omega^2}} \tag{3}$$

$$\alpha_\vartheta = \arctan\left( \frac{k_\vartheta + mgz_S - J_\Theta^S \Omega^2}{c_\vartheta \Omega} \right)$$

With the constant geometry of the vibrator and for the same vibrating depth (same surrounding stresses) the model parameters $k$, $c$, $k_\vartheta$ and $c_\vartheta$ depend only on the density of the surrounding soil. Values for the spring stiffnesses and the dashpot constants for foundation oscillations can be found in (Das 1983; Holzlöhner 1986; Gazetas 1991; Wolf 1994).

## 3.3 Rotation

In the previous sections the vibrator motion was regarded as projection of the rotational motion into a coordinate plane (two-dimensional). The three-dimensional motion can be described by two projections in perpendicular planes. E.g. the motion of the gravity center is:

$$x_S(t) = X_S \sin(\Omega t + \alpha_x), \quad y_S(t) = -Y_S \cos(\Omega t + \alpha_y)$$

If the motion is circular, i.e. $X_S = Y_S$ and $\alpha_x = \alpha_y = \alpha$, a geometrical phase angle of the eccentric mass $\varphi = \pi/2 - \alpha$ can be defined. The eccentric mass precedes the motion of the vibrator by this angle, which lies between $\varphi = 0$ and $\varphi = \pi$.

## 3.4 Measurement of the motion

The horizontal accelerations are measured at the tip and at the shoulder in two orthogonal directions (sensors 1, 2, 3, 4 in Figure 6). Herefrom the amplitudes can be obtained by integration. Further the phase angle of the eccentric mass has to be determined. This can be done by a pulse given by an emitter (sensor 5) when the eccentric mass passes a certain point of the vibrator tube. From the time lag of this pulse to the

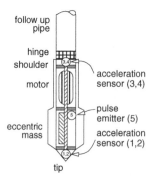

follow up pipe

hinge

shoulder

motor

acceleration sensor (3,4)

pulse emitter (5)

eccentric mass

acceleration sensor (1,2)

tip

Figure 6. Sensors.

zero points of the time history of the accelerations the phase angle can be calculated.

In order to be able to consider a possible influence of the temperature on the acceleration measurement, the temperature of the sensors (1, 2, 3, 4) are measured. The actual depth of the vibrator is recorded, and the hydraulic parameters – oil pressure and flow rate – are measured.

All signals are digitized and analyzed in a laptop. Measuring shots are stored regularly, and some results are displayed on-line.

### 3.5 Evaluation of the measurements

Here the determination of the model parameters, stiffness and damping, is outlined briefly.

### 3.5.1 Calculation of the oscillation

First the motion of the sensors is determined by time integration of the acceleration signals. This is done separately for two coordinate directions. In the sequel the equations are only written for the $x$-direction. The equations for the $y$-direction look similar, otherwise they are separately specified.

From the oscillations of the sensors the motion of each point can be determined by linear interpolation under the assumption of a rigid vibrator, e.g. the motion of the gravity center:

$$x_S(t) = \frac{x_{ti}(t) - x_{sh}(t)}{z_{ti} - z_{sh}}(z_S - z_{sh}) + x_{sh}(t) \,,$$

wherein $z_{ti}$ and $z_{sh}$ are the vertical position, as well as $x_{ti}(t)$ and $x_{sh}(t)$ the motions of the tip and the shoulder sensor, respectively.

The amplitude of the oscillation is determined simply with the maximum values, e.g. for the horizontal oscillation:

$$X_S = \frac{max(x_S(t)) - min(x_S(t))}{2}$$

The parameters of the rocking motion are determined from:

$$\vartheta_x(t) = \arctan\left(\frac{x_{ti}(t) - x_{sh}(t)}{z_{ti} - z_{sh}}\right)$$

$$\Theta_x = \frac{max(\vartheta_x(t)) - min(\vartheta_x(t))}{2}$$

### 3.5.2 Determination of the phase angles

The pulse emitter on the positive $y$-axis gives a signal, when a screw in the eccentric mass passes over it. This results in a square wave signal. The phase angle of the horizontal motion $\alpha_{S,y}$ can be calculated then from the time shift $t_y$ between the middle of square wave signal and the positive amplitude maximum $max(y_S(t))$

$$\alpha_{S,y} = \pi/2 - \Omega t_y \,,$$

for $\alpha_{S,x}$ a rotation by 90° must be considered:

$$\alpha_{S,x} = 3\pi/4 - \Omega t_x$$

For the rocking oscillation similar equations applies.

### 3.5.3 Calculation of the model constants $k$ and $c$

The model constants can be ascertained with the help of the equations of motion (2) and (3). From the horizontal oscillation follows:

$$c_x = \frac{m_u r\Omega}{X_S}\cos\alpha_{S,x} \,, \quad k_x = \Omega^2\left(\frac{m_u r}{X_S}\sin\alpha_{S,x} + m_x\right)$$

From the rocking oscillation follows:

$$c_{\vartheta,x} = \frac{(z_F - z_S)m_u r\Omega}{\Theta_x}\cos\alpha_{\vartheta,x}$$

$$k_{\vartheta,x} = \frac{(z_F - z_S)m_u r\Omega}{\Theta_x}\sin\alpha_{\vartheta,x} - m_x g z_S + J^S_{\Theta,x}$$

Therein the resonant mass of the follow up pipes is considered by $m_x$ and $J^S_{\Theta,x}$. They are determined in such a way that the spring stiffnesses $k$ vanishes for oscillation in air

$$m_x = \frac{m_u r}{X_{S,L}} \,, \quad J^S_{\Theta,x} = \frac{m_u r(z_F - z_S)}{\Theta_{x,L}} + \frac{m g z_S}{\Omega^2} \,.$$

### 4 FULL-SCALE EXPERIMENTS

For these tests a new vibrator of the type Tr 17 was built, in which the sensors (Fig. 6) are internally installed. The sensor signals are transferred by an analogue-digital converter into a laptop. The hydraulic motor of the vibrator is operated with a maximum oil pressure of 300 bar.

118

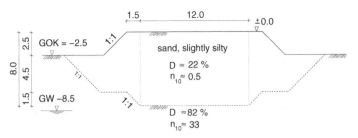

Figure 7. Test pit.

## 4.1 Site conditions

The tests were done in a sand pit in Westernbach, Germany. The soil is a slightly silty sand, with approximately 8% silt. The maximum porosity is in average $n_{max} = 52.3$ in average, the minimum is $n_{min} = 35.3$. The grain density is $\varrho_s = 2.68$ g/cm$^3$.

In order to vibrate in a loose and homogeneous sand, a pit was excavated with 5.5 m depth, and loosely refilled (Fig. 8). With the rest of the excavated sand a 2.5 m embankment was constructed (Fig. 7), which indicates clearly the loosening of the material. As horizontal distance of 4 m between the vibrating points was chosen to minimize a possible interaction.

Figure 8. Refilling.

The density of the in situ soil was determined in two places with the water displacement method. The in situ soil has an average porosity of $n = 38.3$ and thus a relative density $D = I_n = \frac{n_{max} - n}{n_{max} - n_{min}} = 82\%$.

The density of the loosened sand was determined in two depths at five locations each, by placing a bucket on the actual surface during refilling. The density of the material in the bucket is then equal to the density of the surrounding soil. The refilled soil has an average porosity of $n = 48.5$ and thus a relative density $D = 22\%$.

After refilling standard penetration tests with a 50 kg hammer and 15 cm$^2$ tip cross area in the points of the density measurements showed a loose homogeneous refilling. The average number of blows per 10 cm were $n_{10} \approx 0.5$ in the refilling and $n_{10} \approx 33$ in the virgin soil below 8 m depth.

## 4.2 Compaction tests

Preliminary tests using the in situ sand as backfill material showed that compaction with water flushing performed very poor. Therefore attempts without flushing and with air flushing were carried out. Various methods of compaction were tested.

The inflow of the backfill was the main criterion for controlling. In tests with stepwise withdrawing, the vibrator was hold in constant depth until the inflow of backfill abated. In tests with cyclic withdrawing, the cycles were repeated as long as backfill was inflowing.

### 4.2.1 Oscillation of the vibrator

The motion of the vibrator in air is almost circular. The acceleration signals are nearly pure sine functions without harmonic waves. The hinge remains almost in rest.

When the vibrator sinks into the soil the acceleration signals become distorted. The rotational motion becomes elliptic, whereby the amplitude decreases more strongly perpendicular to the fins than in fin direction. The ratio of the harmonics increases. Every point of the vibrator is in motion, also the hinge.

### 4.2.2 Detection of the dense in situ soil

The boundary between the loosened and virgin soil was clearly detected with the measurements in all cases. E.g. in the test without flushing in Figure 9. When reached the virgin soil in 7.7 m depth (after approximately 2.3 min) the vibrator was not able to sink anymore. The oil pressure reached already its maximum at 2.5 m and thus did not indicate the layer boundary, whereas the model constants, spring stiffness and damping, clearly change at the layer boundary. As mentioned before the amplitude decreases more strongly perpendicular to the fins ($x$-direction). Thus the model parameters are represented only for this direction.

### 4.2.3 Detection of compaction end

If the vibrator is held in a constant depth until no more backfill material inflows, all measured and cal-

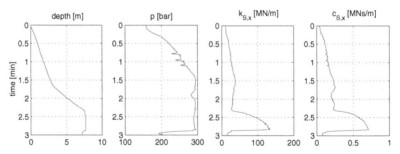

Figure 9. Sinking of the vibrator without flushing.

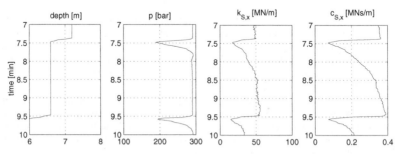

Figure 10. Holding in constant depth, withdrawing when the inflow of backfill abates (9.5 min) - end of compaction.

culated quantities change. If the inflow of backfill abates, it can be assumed that no further compaction takes place. Thus this time is the very end of the compaction.

The oil pressure remains already constant long before this end time (Fig. 10). The mechanical model constants ($k$, $c$) change during the whole compaction. They visualize thereby the compaction process better than the oil pressure. A possible criterion for withdrawing in the next depth level should be based on the mechanical quantities.

During compaction in cycles the calculated stiffness usually increases with each cycle in the same depth (Fig. 11), whereas the oil pressure always achieves the same value. Also this could be used as criterion to stop cycling in a certain depth range.

### 4.2.4 *Resonance*

The soil-vibrator system can oscillate in resonance, with amplitudes larger than those in air and a phase angle $\alpha \approx 90°$. This occurred e.g. during withdrawing in cycles with air flushing. Therefore a single measurement of the amplitude or the phase angle cannot generally be used for compaction control, unless the vibrator frequency is far away from the natural frequency of the system.

The stiffnesses and the damping constants calculated from the amplitude and the phase angle are independent of the ratio between operating and natural frequency and give therefore a more reliable information about the soil.

### 4.2.5 *Correlation with SPT-test results*

A plot of the measurements over the depth can be compared with the results of the SPT-test after compaction 0.75 m away from the compaction center. This plot shows a good qualitative agreement of the stiffnesses (regarded as envelope) with the number of blows (Figs 12, 13). The oil pressure is not comparable with the SPT-tests, it becomes always maximum.

## 5 CONCLUSIONS

An on-line compaction control in deep vibratory compaction was presented. Control takes place via monitoring the vibrator motion. The measured quantities are the accelerations of the vibrator and the temporal location of the eccentric mass. Out of this the model constants of a simplified oscillator are computed, namely the stiffness and the damping. These quantities give some information about the ongoing compaction process.

The motion of the vibrator changed measurably during compaction in all tests. The most important conclusions from the test are:

- The end of the compaction in constant depth can be detected by the mechanical quantities stiffness and damping, and not by the hydraulic

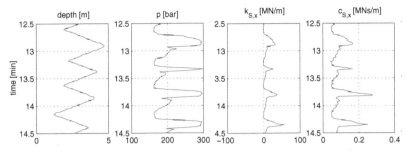

Figure 11. Compaction in cycles with air flushing: Increasing of the stiffness and the damping in each cycle.

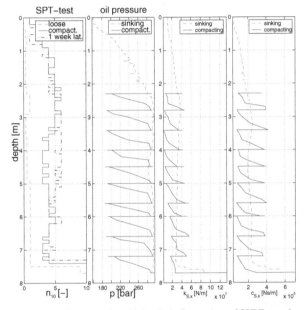

Figure 12. Compaction without flushing (stepwise withdrawing): Comparison of SPT-Test and on-line measurements.

quantities oil pressure and flow rate or the power consumption.

- During compaction in cycles the stiffness usually increases with each cycle in the same depth range.

- The model stiffness is qualitatively comparable to the SPT-test results after compaction.

- Measuring the amplitude or the phase angle alone is only sufficient for on-line compaction control, when the system is rather far away from resonance.

Further test are necessary in order to define precise criteria for the holding time of vibrating, as well as to quantify the correlation between the model constants and SPT-test results.

*Acknowledgement*

The full-scale tests were financially supported by: *Bayerisches Staatsministerium für Wirtschaft, Verkehr und Technologie, Innovationsberatungsstelle Südbayern.*

121

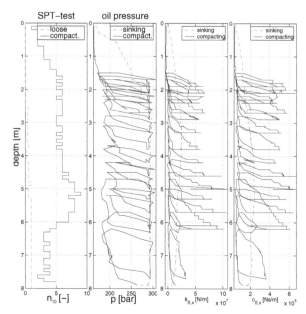

Figure 13. Compaction with air flushing in cycles: Comparison of SPT-Test and on-line measurements.

## References

Broms, B. (1991). Deep compaction of granular soils. In H.-Y. Fang (Ed.), *Foundation Engineering Handbook* (2 ed.)., Chapter 23, pp. 814–832. Chapman & Hall.

Brown, R. (1977). Vibroflotation compaction of cohesionless soils. *Journal of the Geotechnical Engineering Division, ASCE 103*, 1437–1451.

D'Appolonia, E. (1953). Loose sands - their compaction by vibroflotation. *American Society for Testing Materials, Special Technical Publication No. 156*, 138–154.

Das, B. (1983). *Fundamentals of Soil Dynamics*. Elsevier Science Publishing. ISBN 0-444-00705-9.

Fellin, W. (2000a). On-line Verdichtungskontrolle bei der Rütteldruckverdichtung. *Bauingenieur 75*, 607–612.

Fellin, W. (2000b). Quality control in deep vibrocompaction. In D. Kolymbas and W. Fellin (Eds.), *Compaction of Soils Granulates and Powders*, Number 3 in Advances in Geotechnical Engineering and Tunnelling. Balkema.

Fellin, W. (2000c). *Rütteldruckverdichtung als plastodynamisches Problem*. Number 2 in Advances in Geotechnical Engineering and Tunnelling. Balkema.

FG Straßenwesen (1979). Merkblatt für Untergrundverbesserung durch Tiefenrüttler. Forschungsgesellschaft Straßenwesen, Köln: Straßenbau AZ. Sammlung technischer Regelwerke und amtlicher Bestimmungen für das Straßenwesen. Stand Oktober 1984, Abschnitt Untergrundverbesserung.

Gazetas, G. (1991). Foundation vibration. In H.-Y. Fang (Ed.), *Foundation Engineering Handbook* (2 ed.)., Chapter 15, pp. 553–593. Chapman & Hall.

Holzlöhner, U. (1986). Schwingungen von Fundamenten. In W. Haupt (Ed.), *Bodendynamik, Grundlagen und Anwendung*, Chapter 5, pp. 141–188. Friedrich Viehweg & Sohn, ISBN 3-528-08878-8.

Poteur, M. (1968). *Beitrag zur Untersuchung des Verhaltens von Böden unter dem Einfluß von Tauchrüttlern*. Ph. D. thesis, Fakultät für Bauwesen der technischen Hochschule München.

Smoltcyk, U. (2001). *Grundbau–Taschenbuch* (6 ed.), Volume 2. Berlin: Ernst & Sohn.

Wolf, J. (1994). *Foundation Vibration Analysis Using Simple Physical Models*. PTR Prentice-Hall, ISBN 0-13-010711-5.

*Vibratory Pile Driving and Deep Soil Compaction - TRANSVIB2002,*
*Holeyman, VandenBerghe & Charue (eds.), © 2002 Swets & Zeitlinger, Lisse, ISBN 90 5809 521 5*

# Foundation of wind power station on mining dump applying deep soil compaction

Dr.-Ing. Yasser El-Mossallamy
*Associated Prof., Ain Shams University, Cairo, Egypt*
*c/o ARCADIS Consult, Berliner Allee 6, D - 64295 Darmstadt, Germany*

Dipl.-Ing. Jürgen Löschner, Dipl.-Ing. Wolfgang Kissel
*ARCADIS Consult, Berliner Allee 6, D - 64295 Darmstadt, Germany*

and Dipl.-Ing. Bodo Schlesinger
*ARCADIS Consult, Glück-Auf-Straße 1, D-09599 Freiberg, Germany*

ABSTRACT: The wind power station Klettwitz in Germany was constructed on a 90 m thick mining dump. To enhance the foundation performance regarding bearing capacity and deformation behavior, a special soil improvement system applying a combined system of vibro-floatation and vibro-displacement was developed and implemented. With this system the tower of the wind power turbine can be founded on shallow footing fulfilling both the ultimate limit and the serviceability limit requirements. Numerical analyses were carried out during the design stage to study the performance of the foundation and to prove the design criteria. The results of the numerical analyses, the in-situ tests such as CPT, which were executed before and after conducting the soil improvement measures, as well as the settlement measurements will be presented and discussed. The results demonstrate the efficiency of this combined soil improvement system to enhance the performance of sensitive structures on inhomogeneous, subsidence-prone mining dumps.

## 1 INTRODUCTION

Wind power has become popular all over the world as a renewable "green energy". Following the Kyoto Conference on Global Warming in 1998, German authorities and utilities declared environmentally-friendly energy a priority and decided to promote wind energy.

The wind power station Klettwitz represents an important step in the realization of the recommendations of the Kyoto Conference. The generated power will be sufficient to supply more than 31000 family houses with environmentally-friendly electricity. The wind power station is not only a technical achievement but also offers economic advantages.

This paper deals with aspects concerning design, performance and quality control of the foundations of the wind power station.

## 2 THE WIND POWER STATION AND THE DESIGN REQUIREMENTS

The wind power station Klettwitz in Germany is the first and largest wind-powered electricity-generating plant to be constructed on a 90 m thick mining dump.

The power station covers an area of about 228 ha. It consists of 38 wind turbines Type Vestas V 66 – 1.65 MW with rotors 66 m in diameter (Fig. 1). The rotor is installed on a 78 m high steel tower with a diameter of 4.0 m at its base and 2.3 m at the top.

The minimum distance between the towers is about 200 m.

Figure 1. Layout of the wind power station.

The own weight of the super-structure reaches about 2.2 MN. The moment at the tower base in case of maximum working conditions (operational wind) reaches 14.7 MNm. In case of extreme load conditions the maximum moment at the tower base reaches about 35.9 MNm. This load case corresponds to a wind velocity of more than 180 km/h (max. wind velocity through 50 years), when the turbine should be automatically turned off.

The design of the foundation had to fulfill the following serviceability requirements:

- The required range of the natural frequency of the whole structure is 0.4 Herz to ensure optimum operation conditions. Therefore, a massive foundation was required.
- Foundation tilting had to be smaller than 1:125.

## 3 GEOTECHNICAL AND HYDROLOGICAL CONDITIONS

The Klettwitz wind power station is located in Niederlausitz, about 60 km north of Dresden in Germany. The power station was designed to be erected on a coal mining dump originating from the years 1951 to 1991. The subground consists of fill (dump materials) about 90.0 m in thickness followed by natural tertiary clay soils and fine sand. The dump consists mainly of loose deposits of relatively large heterogeneity regarding both particle size and density. Figure 2 shows the grain-size distribution of the dump materials.

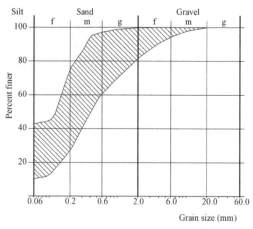

Figure 2. Grain size distribution of the dumping materials.

A traditional soil exploration as well as an extensive in-situ investigation program were conducted to obtain a reliable estimation of the soil parameters. A large number of cone penetration tests (CPT) were carried out to investigate the in-situ conditions. The results of the CPT show the high heterogeneity of the dump materials. The results of the CPT were evaluated according to the German standards. Regarding the CPT results, the dump materials can be divided into three domains as follows:

- Upper domain (20 to 25 m thick) with average cone resistance less than 2.5 MPa, i.e. very loose materials
- Intermediate domain (about 15 m thick) with cone resistance between 5.0 to 7.5 MPa, i.e. loose to median dense materials
- Lower domain with cone resistance between 7.5 and 15 MPa, i.e. median dense materials.

These three domains correspond very well with the applied dumping technique and dump materials. Figure 3 shows typical results of the conducted CPT tests.

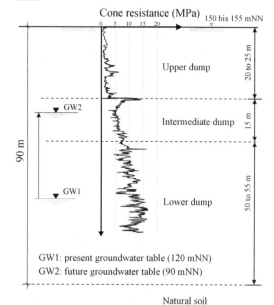

GW1: present groundwater table (120 mNN)
GW2: future groundwater table (90 mNN)

Figure 3. Subground stratification and CPT results.

The deformation modulus of the different dump layers were determined using the equation

$$E_s = \alpha * q_c \qquad (1)$$

where $E_s$ = deformation modulus; $\alpha$ = dimensionless factor; and $q_c$ = cone resistance.

The $\alpha$ factor depends on the soil type. A value of $\alpha$= 2.0 was found reasonable and led to results in good agreement with the results of the oedometer tests.

Due to neighboring mining activities and the required dewatering, the groundwater table currently lies about 60 to 65 m below the ground surface. Therefore, design had to consider the expected ca. 30 m rise in the groundwater table after the end of the neighboring mining operations. The subsidence index of the dump materials was determined from odometer tests and its value ranges from 0.15 % to 0.4 %.

The soil parameters were determined based on the laboratory and in-situ tests as well as on experience gained in similar projects. These parameters are given in Table 1.

Table 1: Geotechnical parameters

| Soil parameter | | Upper dump | Intermediate dump | Lower dump |
|---|---|---|---|---|
| $\gamma/\gamma'$ | [kN/m³] | 17/10 | 17/10 | 17/10 |
| $E_s$ | [MPa] | 5 | 10 | 10 to 20 |
| $c'$ | [kPa] | 2.5 | 2.5 | 2.5 |
| $\varphi'$ | [°] | 25 | 25 | 25 |

where:

$\gamma/\gamma'$    Total / Effective unit weight of soil
E    Deformation modulus
$c'$    Cohesion and $\varphi'$ Angle of internal friction

## 4 FOUNDATION SYSTEM

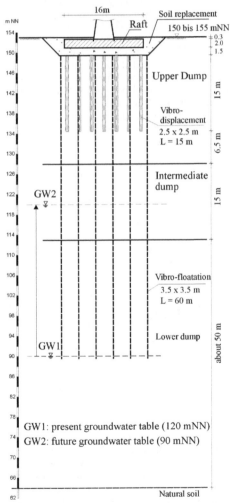

Figure 4. Foundation concept.

To enhance the foundation performance regarding bearing capacity (ultimate limit state) and deformation behavior (serviceability limit state), a special soil improvement system was developed and implemented. Soil improvement consisted of a combined system of

- vibro-floatation down to a depth of about 60 m below ground surface to reduce the soil subsidence following the rise in the groundwater table; the vibro-flotation was carried out with a grid of 3.5 x 3.5 m,
- vibro-displacement with a column length of 15 m and grid spacing of 2.5 x 2.5 m to reduce the settlement of the structure loads and to produce a homogeneous soil matrix beneath the foundation level within its influence depth,
- and a load-distributing compacted layer with a depth of 1.5 m to even out the foundation level.

The superstructure was then built on a relatively massive reinforced concrete raft measuring 16 x 16 m with a thickness of 2.0 m. Figure 4 schematically illustrates the foundation concept.

## 5 NUMERICAL ANALYSES

Two and three-dimensional finite element analyses were conducted to model the foundation behavior under various load cases. The analyses were carried out in the following stages:

Stage 1: The first step of the analyses was to estimate the improvement factor of the vibro-displacement columns. The improvement factor is defined as the settlement without stabilized columns divided by the settlement with stabilized columns related to only the height of the improved layer. An axisymmetric finite element model was applied to analyze the performance of a large extended forest of vibro-displacement columns (Fig. 5).

The aim of this model is to estimate the improvement factor. Isoparametric triangular elements with 15 nodes (Program PLAXIS) were applied in this analysis. Soil and material of the vibro-displacement columns were idealized using the Mohr-Coulomb model.

The applied soil parameters are given in Table 1. The coarse-grained materials used for the construction of the stabilized columns consist of gravel with an internal friction angle of 35°. The calculated improvement factor reaches about 1.4. This value agrees well with the calculation value according to the charts of Priebe (Priebe 1995).

Stage 2: The behavior of the foundation under various load cases was then simulated using a three-dimensional boundary element analysis taking into account stiffness of the raft, soil improvement due to the construction of the vibro-displacement columns, and the different soil stratification.

The maximum working conditions (operational wind) including the load eccentricity were considered in these analyses. The calculated settlement ranged from 10.0 to 18.0 cm.

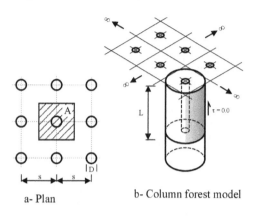

a- Plan

b- Column forest model

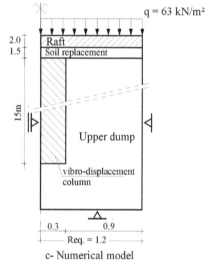

$q = 63$ kN/m²

c- Numerical model

Figure 5. Column forest model.

Stage 3: In case of extreme load conditions (max. wind velocity through 50 years, when the turbine should be automatically turned off), the calculations were performed applying a larger deformation modulus of the soil (three times the static values) to consider the dynamic effect. The additional differential settlement and tilting due to this load case were then determined. The additional settlement under these extreme load conditions reaches about ± 1.0 cm.

Stage 4: Differential settlement within the foundation due to soil subsidence in the improved soil layers beneath the foundation as a result of the rise of the groundwater table is minor and can be ignored. This is due to the vibro-floatation and vibro-displacement columns beneath the foundation. Nevertheless, the effect of non-equal soil subsidence in the close vicinity of the foundation had to be investi-

gated. This effect was studied using a two-dimensional finite element analysis (see Fig. 6).

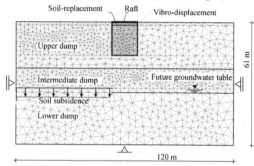

Figure 6. FEM-Model to study the effect of non-equal soil subsidence.

After conducting all analyses, the serviceability requirements of the foundation regarding its tilting were checked. Figure 7 shows the results of the calculated tilting depending on the amount of differential soil subsidence in the close vicinity of the foundation. It can be seen that the tilting limit can be achieved with a maximum differential subsidence of 15 cm. The 15 cm subsidence corresponds to a 30 m water rise with a subsidence index of 0.5%. This value of the subsidence index is larger than the values determined in the laboratory tests and agrees well with experience.

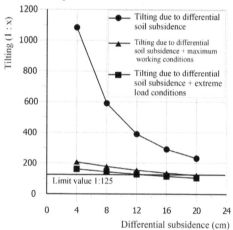

Figure 7. Relationship between foundation tilting and differential soil subsidence.

The bearing capacity of the foundation was checked according to the German standards. The internal capacity of the individual vibro-displacement columns was checked according to Brauns (1978).

The distribution of the subgrade reaction modulus beneath the raft foundation was determined from the results of the 3-D analyses. The structural engineer then based the design of the foundation and the

126

tower on the values of the determined subgrade reaction modulus. The subgrade reaction modulus determined by the enhanced geotechnical analyses can be considered as a comfortable interface between the geotechnical and the structural engineers for adequately considering the soil-structure-interaction.

# 6 CONSTRUCTION FEATURES

The installation principles of vibro-floatation and vibro-displacement are well known. This paper only focuses on those features, that are important for construction productivity, quality control and quality assurance.

The vibro-flotation technique is based on the fact that soil particles can be rearranged into a denser condition under the influence of vibrations. The depth of vibro-floatation reached about 65 m below the ground surface with a total number of 36 points installed beneath each foundation of the wind power station. With vibro-displacement, the vibrator penetrates the weak soil to the design depth and the resulting cavity is filled with gravel material, which is compacted in place. Forty-nine columns of a length of about 15 m are installed beneath each foundation.

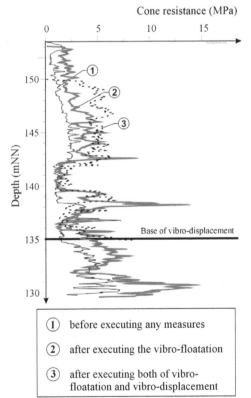

① before executing any measures

② after executing the vibro-floatation

③ after executing both of vibro-floatation and vibro-displacement

Figure 8. Comparison of CPT results before and after executing the soil improvement measures.

The installation parameters are recorded as part of the standard quality assurance procedure in order to verify the reached depth and degree of compaction. The depth, the applied energy, the penetration and the withdrawal speed of the poker vibrator are automatically recorded for each installation point. By vibro-displacement, during withdrawal of the vibrator and in addition to the above-mentioned recording, material quantities were recorded to ensure the column integrity and to prove that a minimum area replacement ratio was maintained. The measurements show the applied gravel quantity of 0.5 to 1.5 ton/m. This ensured a minimum column diameter of 0.6 m.

A relatively high productivity of about 30 linear meters per rig and hour was achieved by simultaneous installation of the vibro-flotation and vibro-displacement columns.

To verify the soil improvement, CPT tests were carried out after execution of vibro-flotation and vibro-displacement and compared with the results of the CPT before these measures. Figure 8 shows an example of this comparison.

# 7 MONITORING PROGRAM

An extensive measurement program was performed to verify the design concept, to ensure the serviceability requirements and to fulfill the quality control. This included leveling measurements on the foundation at its four edges, measurements of the tilting of the tower immediately after its construction, leveling measurements on the dumping fill and measurements of groundwater conditions.

The maximum measured settlement after erecting the power station amounts to about 8.0 cm. The immediate settlement due to the foundation and structure weight was not measured and estimated to be about 2 to 4 cm. Therefore the total settlement reaches about 10 to 12 cm and agrees well with the calculated value. The maximum foundation tilting is about 1:300. These measurements show that all design criteria are fulfilled.

The groundwater measurements show no remarkable change in the groundwater table.

# 8 CONCLUSIONS

A combined soil-improvement system including both vibro-flotation and vibro-displacement techniques has proved its ability as a very suitable system to construct sensitive structures such as a wind power station on a heterogeneous dumping fill. The technical reliability and economical merits have been shown.

The particular advantages, namely high productivity, low mobilization costs, high flexibility in design and application, and potential use in highly heterogeneous soils make it increasingly interesting for similar development and infrastructure projects on dumping fills.

Different design procedures allow a wide spectrum of application of this combined system for improving the deformation and stability characteristics of weak soils extending to a large depth with a high level of confidence.

## 9 REFERENCES

Balam, N.P. and Booker,J.R. 1985. Effect of stone columns yield on settlement of rigid foundations in stabilized clay, *Int. Journ. For Num. And Anal. Meth. In Geomec. Vol 9, 331-451.*

Brauns, J. 1978. Die Anfangstraglast von Schottersäulen im bindigen Untergrund. *Die Bautechnik 8/1978, 263-271.*

Lunne, T., Robertson, P.K. and Powell, J.J.M. 1997. Cone penetration testing in geotechnical practice. Published by Balckie academic and professional, London

PLAXIS 1999. *Finite element code for soil and rock analysis.* Version 7.11, Manual, Plaxis B.V, Balkema.

Priebe, H. 1995. Die Bemessung von Rüttelstopfverdichtungen *Bautechnik 72/1995,Heft 3, 183-191.*

*Vibratory Pile Driving and Deep Soil Compaction - TRANSVIB2002,*
*Holeyman, VandenBerghe & Charue (eds.), © 2002 Swets & Zeitlinger, Lisse, ISBN 90 5809 521 5*

# On- and Offshore Vibro-Compaction for an oil pipeline in Singapore

W.C.S. Wehr
*Keller Grundbau GmbH, Offenbach, Germany*

V. R. Raju
*Keller Foundation (S.E. Asia) Pte Ltd., Singapore*

ABSTRACT: Extensive ground improvement using vibro compaction of reclaimed sand fill was carried out below a future twin crude oil pipeline on Jurong island in Singapore. Compaction was executed to depths ranging between 20m and 30m. Onshore compaction was done underneath the future pipeline and offshore compaction on the seaside slope at the VLCC (Very Large Crude Carrier) jetty. In total, over 1.9 million cubic meters were compacted over a 3 month period.

## 1 INTRODUCTION

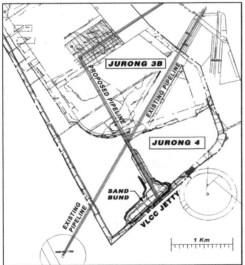

Figure 1. Plan drawing showing site layout

The construction of an approx. 7 km long twin crude oil pipeline on reclaimed land at Jurong Island required extensive ground improvement using vibro compaction. The compaction was carried out to limit future settlements of the sandfill directly underneath the pipeline and to stabilize the seaside slope of the sand bund at the pipeline jetty. This paper describes the technical requirements of the works, the practical aspects of execution of the works and presents post compaction cone penetration test results.

Figure 1 shows a plan layout of the site and the location of the existing and proposed pipeline and the VLCC jetty. Onshore compaction work was carried out by Keller for a 4.5 km length of the pipeline on the Jurong 3B area and offshore compaction over a 1000 m length of the sandbund at the VLCC jetty area.

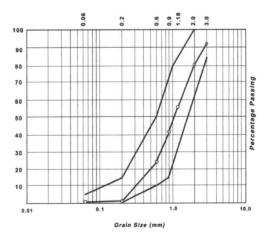

Figure 2. Typical grain size distribution curves of reclaimed sand at Jurong Island

## 2 SOIL CONDITIONS

Generally the reclaimed soil is made up of medium to coarse sand with the grain size distribution curves of figure 2. The silt content was less than 5% and well suited for compaction. The ground water level is situated at a level of +1.0 ACD approximately 5m below the ground surface.

## 3 VIBRO COMPACTION

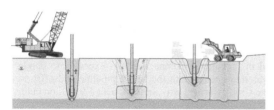

Figure 3. Sequence of steps for vibro-compaction – penetration to full depth and compaction during withdrawal in steps

Vibro compaction (also referred to as vibroflotation) is a technique used to densify granular soils using the depth vibrator. Horizontal vibrations are produced by a rotating eccentric weight which is driven by an electric motor mounted within the vibrator. Depending on the speed of rotation of the eccentric weight, the weight of the vibrator and the eccentric weight and the geometry of the vibrator, centrifugal forces in the range of 200 to 350 kN can be developed. Vibration amplitudes range between 20mm and 35mm.

The vibrator with its motor running is lowered into the ground with the assistance of water jets (see Figure 3). Once the vibrator has reached the required depth, the water jets are turned off and the vibrator is pulled back in short steps (typically 40cm – 60cm). At each step, the vibrator is held in position for a predetermined period of time (typically 30 – 50 seconds). During this period, the inter particular friction is temporarily reduced and the sand particles are rearranged in a denser state. The process is repeated until the required depth of sand is compacted, leaving behind a column of compacted sand. The vibrator is then positioned at the next compaction point and the process is repeated. Compaction points are typically arranged either in a square or equilateral triangular grid pattern.

A benefit of using depth vibrators is that the source of energy (the motor) is mounted within the vibrator which penetrates the ground and imparts the energy directly at the point of compaction. This results in a very efficient compaction process with minimum loss of energy. The other advantage is that only selected layers can be compacted (if so desired). For example, as in the case of this project, only the bottom 14m can be compacted leaving the top 6m uncompacted (refer to section 4).

## 4 COMPACTION UNDERNEATH PIPELINE

The foundation design for the pipeline called for the densification of a 20m wide strip of the loose reclaimed sand underneath the pipeline to a relative density of 70% (see Figure 4). It was not necessary to compact the layer above the pipeline (between level +6.0m ACD and –1.5m ACD) as this was to be excavated out anyway.

Prior to compaction, pre-CPT tests were carried out in each 20m wide and 50m long panel. A representative example is given in figure 5. Precompaction cone resistances varied between 5 MPa and 8 MPa (1 MPa = 10 kgf/cm2). Friction ratios were generally about 0.5%.

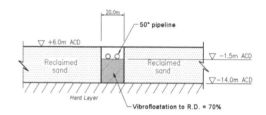

Figure 4. Cross section showing pipeline and treatment area

Based on the results of a field trial with varying compaction point spacings, an equilateral triangular grid with 3.5m spacing was chosen for compaction to 70% relative density underneath the pipeline. The settlement of the sand surface was approximately 1.5m. Following compaction in each 50m x 20m panel, one post-CPT test was carried out in the centroid of the triangle formed by the compaction points. Figure 6 shows a typical post compaction CPT result. The required tip resistance corresponding to 70% relative density based on a correlation by Schmertmann (1978) is also shown.

## 5 COMPACTION IN SAND BUND

At the sand bund, compaction was carried out to 70% RD in a 20m wide strip below the pipelines over the full depth of the sandfill as the pipeline was now located at the top of the sandfill, see figure 7.

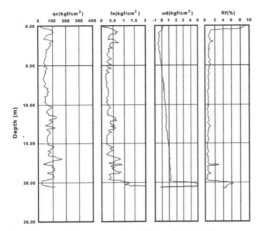

Figure 5. Typical pre compaction CPT result in pipeline area

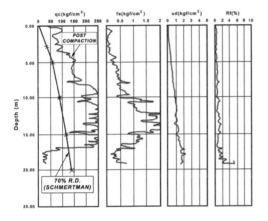

Figure 6. Typical post compaction CPT result and tip resistance corresponding to 70% relative density

The average compaction depth was 27m below the ground. Adjacent to this strip the 1:3 inclined slopes had to be densified to the JTC specification which corresponds to a tip resistance of 4, 6, 8 and 10 MPa below a depth of 0m, 2m, 8m and 24m below ground level.

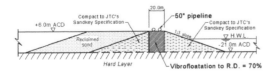

Figure 7. Cross section of sand bund showing treatment area corresponding to 70% relative density

A triangular grid with 3.5m spacing was chosen below the pipelines to achieve the 70% relative density requirement. Compaction was carried out in a manner similar to that described in section 4 above. A triangular grid with 4.0m spacing was chosen in the sea side slope area to achieve the JTC specification. Compaction was carried out using barges (see figure 8). The positioning of the barge-based vibrator units was realized with a GPS-system having an antenna on top of the crane boom and a fixed land station to provide a stable signal.

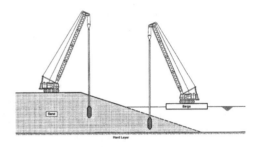

Figure 8. Schematic cross section showing on and offshore compaction rigs

In every second pipeline and slope panel one Post-CPT test was carried out in the centroid of the triangle formed by the compaction points. The CPT's on land were executed with a conventional CPT machine and the CPT's in the sea were performed with a special CPT tower which was transported with a barge and positioned by divers. A typical post CPT result is shown in figure 10, showing as well the required cone resistance by the JTC specifications with four different specified cone resistances in four different depths.

6  VIBRATION MEASUREMENTS

In one panel of the crude oil pipeline area a gas pipeline had to be crossed. To ensure that the existing gas pipeline which is situated approximately 2m-3m below ground level is not damaged, vibration measurements were carried out on the ground surface directly above the pipeline. A low frequency vibration meter RION VM81 was used with 3 input channels measuring the vertical particle velocities of the soil at the surface by means of geophones. During sand compaction using three crane vibrator units being 20m away from the gas pipeline, maximum velocities of only 3mm/s were recorded.

This value lies well below the recommended maximum value according to the German standard, DIN 4150, of Vmax = 20mm/s - 40mm/s for industrial structures at frequencies between 10Hz-50Hz.

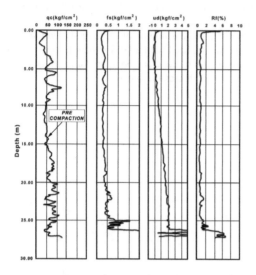

Figure 9. Typical pre compaction CPT result in sand bund area

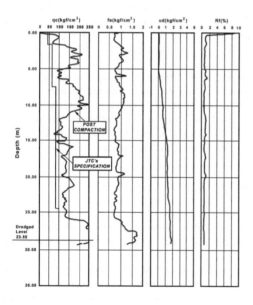

Figure 10. Post CPT with JTC specifications

# 7 CONCLUSION

Extensive ground improvement using on shore vibro compaction of over 1.2 million $m^3$ was carried out below a future crude oil pipeline on Jurong island in Singapore and over 0.7 million $m^3$ of offshore compaction on the seaside slope at the VLCC jetty. Three crane hung vibrator units were employed for the offshore works . To compact the seaside slope, two 150ton cranes including GPS systems were mounted on barges. All five units managed to meet the tight time schedule of only 3 months.

The requirement of 70% relative density below the pipeline and the JTC specification underneath the slope was achieved without any major problems. The magnitudes of vibrations were measured while the compaction works were close to an existing pipeline. Only small ground velocities below 3mm/s were recorded when 3 vibrator units were working about 20m away.

# 8 REFERENCES

Schmertmann, J.H. 1978. Study of feasibility of using Wissa-type piezometer probe to identify liquefaction potential of saturated fine sands. *Technical Report S-78-2, University of Florida.*

Jurong Town Council (JTC) 1999. Civil works at Jurong Island, Singapore, *Technical Specifications, Addendum No.2, Main contract for soil improvement works*

*Environmental aspects of
vibratory operations*

# A case study on safe sheet pile driving with vibration monitoring

W. Haegeman
*Soil Mechanics Department, Ghent University, Belgium*

ABSTRACT: For the rehabilitation and stability improvement of a dike sheet piles were vibratory driven into the crest of the embankment. However, at the toe of the dike at a depth of 0.8 m a gas pipe had been installed supplying the harbour of Antwerp with liquid hydrogen under a pressure of 100 bar. The horizontal distance to the sheet pile wall over a length of 800 m was approximately 6 m. The vibration levels that could occur at the pipe during driving had to be assessed and despite little knowledge on the level of vibrations that could result in damage to the gas line, the permissible limit values had to be established. Finally during driving of the sheet piles the vibrations and settlements were continuously measured at critical locations. The driving activities were completed without any damage. This paper comments on soil conditions and preliminary vibration measurements in the absence of the gas pipe as well as definition of acceptable vibration levels and finally the monitoring results.

## 1 INTRODUCTION

Ground vibrations produced during pile driving operations often annoy people living around construction sites or those vibrations may have detrimental effects on structures or equipment. Therefore in recent years vibration assessment and monitoring has become one of the popular public and scientific interest. This is even more the case when the structure is a high pressure gas line.

After a dike failure rehabilitation of the canal profile was necessary by installing a sheet pile wall into the crest of both embankments along the total length of the canal. On the south side, over a length of 800 m, a liquid hydrogen pipeline is routed parallel to the canal at the toe of the embankment at a depth of 0.8 m. It was told that a mixture of 4 % hydrogen in air is highly explosive with an invisible burning flame at daylight; so safety of the people and pipeline should be guaranteed.

Figure 1 shows a cross section of the dike with the adjacent pipeline. The height of the embankment is ± 2 m and the toe of the sheet pile wall extends approximately 1 m below the pipeline. Horizontal distance between sheet piles and pipeline varies between 6 and 7.9 m.

Soil boring and cone penetration data in the vicinity of the construction site are given in figure 2. Underneath the soft embankment material a fine sand layer extends to a depth of 8 m. Typical grain size distributions of both materials are shown in figure 3.

Several meetings were held prior to construction to coordinate communication among all groups. It was critical that the pipeline engineering consultants be aware of the timing of construction activities and pipeline response measurements. The following actions were decided upon:
- The distance between pipeline and sheet pile wall should be maximized by shifting the wall from the middle of the dike to the internal edge of the embankment of the canal.
- Sheet piles type PU6, about 4 m long should be driven by an excavator mounted high frequency vibratory hammer.
- The 114.3 mm diameter pipeline should a wall thickness of 4.8 mm and steel quality API 5L GRAD B (equivalent to X 42) under working pressure of 100 Bar, should be located exactly.
- Allowable vibration level of the pipeline should be set.
- Preliminary vibration measurements should be performed in a section of the canal in absence of the pipeline to assess if extra measures are necessary in view of the fixed allowable vibration level.
- Finally, during construction near the pipeline deformations and vibrations should be measured and measurement of the pipeline response should guide construction work of the sheet pile wall.

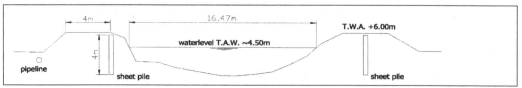

Figure 1. Cross section of the dike

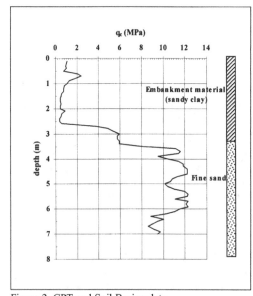

Figure 2. CPT and Soil Boring data

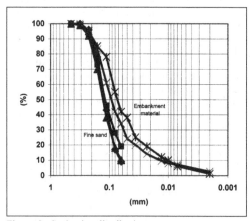

Figure 3. Grain size distributions

## 2 VIBRATION CRITERIA

Factors considered in establishing a vibration criterion were the type and diameter of the pipeline, depths of burial, type of soil, pipeline support conditions and previous experience based on documented reports and journal papers from other projects.

Positive factors in allowing continued high-pressure operation of the pipeline during construction were the type of material, i.e., a ductile steel, as well as quality controls during the fabrication/installation of the pipeline, which included ongoing monitoring/inspection activities, 100 % x-ray of welds and 100 % hydrostatic pressure testing.

Crack susceptibility of piping from blasting vibration indicated that pipelines could withstand vibration levels as high as 150 to 200 mm/s which allow specification of a conservative control limit of 50 mm/s for pile driving vibrations (Dowding, 2000).

However, there can be a significant probability that driving into the underlying sands would produce vibratory densification.

Based on calculations taking into account spherical wave front shape and typical soil, pipeline and blast characteristics, Gantes et al. (2002) reported allowable vibration levels for buried steel pipelines subjected to explosions in the range of 100-160 mm/s.

Prefabricated concrete pile driving and sheet pile driving under similar conditions at 3.5 to 10 m distance from a pressurized pipeline in the harbour of Antwerp caused peak vibration levels at the pipeline of maximum 12 mm/s; reported strains were in the elastic region of the material and no damage was noticed (WTCB, 2000).

Despite these reports and observations, a conservative approach was imposed by the gas company. The guideline was a field criterium for the mid span vibration motion of a free pipe between supports, not taking into account any ground support conditions (Maten, 1984). Due to these considerations a conservative limit of 15.2 mm/s was set for construction induced vibration.

From this literature and project overview it also became apparent that the controlling factor might be pipeline settlement and not vibration. This conclusion is not surprising from a historical viewpoint. There is strong evidence that concern for potential damage to buried steel pipelines during construction should focus on differential movement caused by soil liquefaction, densification, etc. There have been several documented occurrences of such types of failures (O'Rourke and Hall, 1988), but no reported failures because of excessive structural (pipeline) vibration effects.

So, the limiting construction induced stress on the gas pipeline for the project was choosen as that in the longitudinal direction. The primary operating stress in the pipeline in the circumferential direction is a combination of stresses produced by the soil overburden and internal pressure. Stress in the axial direction is the flexural stress induced by differential movement of the ground. Methods were reviewed to estimate the flexural stress induced by the deformed shape of the pipeline caused by settlement. Comparison with limiting construction induced stress allowed for the maximum differential settlements.

However, finally the gas company accepted no permanent deformations along the pipeline.

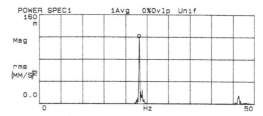

Figure 4. Power spectrum of a vibration registration.

horizontal direction perpendicular to the embankment (channel 2) and horizontal direction parallel to the embankment (channel 3). Driving a sheet pile took about 40 sec with the highest velocity amplitudes at the end of driving.

## 3 MEASUREMENT SYSTEM

Vertical movements of the pipeline were measured by land surveying methods. Ground vibrations were measured with velocity transducers.

Because of the pipeline proximity to the sheet pile wall construction, it was decided to first monitor vibrations at the toe of the embankment on the surface and at a depth of 0.8 m in absence of the pipeline. Several continuous data registrations with a time length of 16 sec, allowing for a frequency analysis, were performed with two Mark Products L-4 1 Hz geophones and a HP 3562A dynamic signal analyzer. Secondly, a three-component velocity vibration transducer used with a IFCO VMS peak reading continuous monitor was installed to record peak vertical and horizontal vibrations over a longer period of time at 0.8 m depth.

Along the pipeline, vibration monitoring positions were choosen initially every 20 m and survey positions were established every 50 m. Vibrations were only recorded with the three component velocity transducer and the peak reading monitor on the soil 20 cm above the pipeline.

The settlement survey measurement of each position consisted of a survey rod positioned on a circular clamp fixed to the pipeline. Vertical permanent settlement was measured with a level monitor.

## 4 MONITORING RESULTS

Sheet pile driving with the vibratory hammer in absence of the pipeline produced the greatest soil response near the operating frequency of 23 Hz.

Figure 4 shows the power spectrum of a vibration registration at 0.8 m depth clearly indicating the frequency peak.

Figure 5 shows a complete record for driving a set of six sheet piles with peak velocity amplitudes every second in vertical direction (channel 1),

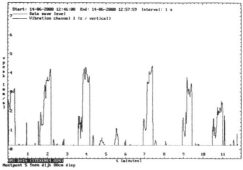

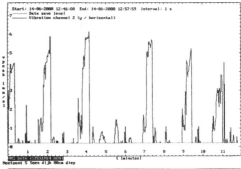

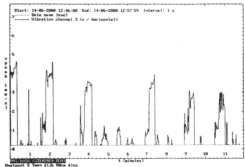

Figure 5. Peak particle velocities in vertical (top) and horizontal (middle and bottom) directions during driving of six sheet piles.

Looking more in detail, there is an initial build-up of response amplitudes as the motion of the vibratory hammer increases in frequency and passes the natural frequency of the soil, followed by a stationary frequency regime with lower response amplitudes, finally increasing to the maximum values because the sheet pile tip approached the vibration monitoring level.

During the ongoing measurements and analysis, it became apparent that the horizontal vibrations perpendicular to the embankment were substantially greater than the velocity amplitudes measured in the other directions. These peak velocities reached a value of 6 mm/s while in vertical and the other horizontal direction 3 to 4 mm/s.

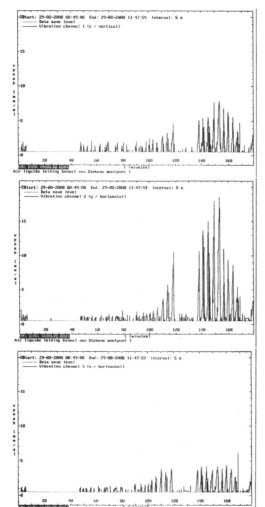

Figure 6. Peak particle velocities in vertical (top) and horizontal (middle and bottom) directions during driving near pipeline.

No noticeable differences could be seen between the registrations at the surface or at 0.8 m depth. All measured vibrations were less than 15.2 mm/s, which was the control established for the campaign so save sheet pile driving near the pipeline could be assumed

However, figure 6 shows the vibration registrations over a period of three hours at the first position on the pipeline while the sheet pile driving is approaching. As can be seen, all components of the vibration are gradually increasing since distance decreases. Vibrations are at maximum for the closest sheet pile (time = 153 min) and afterwards are decreasing again. At that moment the vertical velocity amplitude is 8 mm/s and the horizontal amplitude at channel 2 18.5 mm/s, exceeding the defined vibration criterium.

Construction works were temporary stopped after these three hours and deliberation on the matter was necessary. Reconsidering above mentioned vibration criteria, knowing that the peak velocities only occur over a short period of time at the end of every sheet pile driving and measuring no settlements yet, the gas company allowed for a higher vibration level. It was decided that the horizontal component perpendicular to the embankment never exceeds 20 mm/s, still restricting all settlements to zero.

This control level could be guaranteed during sheet pile installation along the pipeline and construction work resulted in no pipeline settlement nor damage.

The higher vibration velocity amplitudes at the pipeline in relation to the preliminary vibration measurement were attributed to the lower height of the embankment.

5 CONCLUSIONS

A pipeline monitoring program for measuring and evaluating vibrations and settlement due to sheet pile vibratory driving is presented. Due to lack of information and knowledge about buried pipeline behaviour, a first task was to define an acceptable vibration level.

The measurement data appeared to provide a reasonably accurate measure of the pipeline performance during construction and they formed the basis of quantitative decision making. This decision making included an increase of the acceptable vibration level during works.

It is believed that this updated vibration level was still severe in relation to the performance of the pipeline, however at higher vibration amplitudes excessive permanent displacement are more likely to be a far greater concern.

# 6 ACKNOWLEDGEMENT

The monitoring program was paid by the Water Department, Ministry of Flanders, Ghent, Belgium. The writer of this paper gratefully acknowledge and appreciate the cooperation of H. Van Damme, Head of the Department and K. Haelterman, Geotechnics Division, Ministry of Flanders, Ghent, Belgium.

# REFERENCES

Dowding, C.H. 2000. Construction vibrations. *ISBN 0-9644313-1-9.*

Gantes, C.J., Bouckovalas, G.D. & Gerogianni , D.S. 2000. Safety of buried steel pipelines subjected to explosions. *http://cikla.fsv.cvut.cz/~eurostee/085/085.htm.*

Maten, S. 1984. Field criteria for pipe vibration. *Journal Hydrocarbon Processing,* July 1984, pp. 107-108.

O'Rourke, T.D. & Hall, W.J. (1988). Engineering planning and practice for pipeline systems. *Prentice Hall, Englewood Cliffs, NJ.*

WTCB (2000). *Test report DE 611 X814.*

*Vibratory Pile Driving and Deep Soil Compaction - TRANSVIB2002,*
*Holeyman, VandenBerghe & Charue (eds.), © 2002 Swets & Zeitlinger, Lisse, ISBN 90 5809 521 5*

# Settlement due to sheetpile extraction, results of experimental research

P. Meijers
*GeoDelft & Delft University of Technology, Delft, The Netherlands*


A. F. van Tol
*Delft University of Technology, Delft, The Netherlands*

Removal of sheetpiles may cause damage to adjacent buildings. One of the possible causes of damage is settlement of the subsoil due to vibrating. Until now only a few models are available to predict these settlements. In order to get a better understanding of the phenomena involved in the process of settlement during sheetpile extraction a serie of model tests was performed. The test set-up and the first results of analysing these tests will be presented. The results of these tests will be used in the development of a more sophisticated model to predict the settlement during sheetpile extraction.

## 1. INTRODUCTION

Vibrating of sheetpiling, either during installation or during extraction, may be a source of damage to adjacent structures. Well known is damage due to vibrations in the building itself, but also densification of the soil around the sheetpile may cause damage. In the literature several cases are reported. Possible types of damage are:

- damage to buildings due to (uneven) settlement
- damage to buildings due to horizontal deformations
- damage to buildings due to vibrations
- damage to pipelines due to differential settlements or horizontal deformations
- settlement of railway lines
- settlement of existing roads
- instability of structures due to loss of bearing capacity as a consequence of excess pore pressures

The settlement can become quite excessive. In loose sad settlements in the order of 0.5 m close to the sheetpile have been observed. Examples of measured settlements are shown in e.g. (Clough, Chameau 1980), (Lacy, Gould 1985) and (Fujita1994).

Until present only a limited number of models is available to quantify this settlement. All models are empirical of nature and present at best an indication of the settlement to be expected.

In order to fill this gap in knowledge a study was recently started at Delft University of Technology. This study is part of Delft Cluster project DC4 PB3. In Delft Cluster five Delft knowledge institutes, active in the field of civil- and hydraulic engineering, have combined forces. The core of Delft Cluster consists of Delft University of Technology, GeoDelft, IHE, TNO-Building Research and WL|Delft Hydraulics. The collaboration is in the form of an open network, whose aim it is to develop and distribute knowledge in this field.

The aim of project DC4 BP3 is to develop a practical method for the risk assessment of the execution of building project in an urban environment. Part of the risk is damage to adjacent buildings. To develop the overall model use will be made of available knowledge. For parts on which the available knowledge is insufficient additional research will be performed. One of the subjects identified for additional research is settlement during sheetpile extraction. The aim of the research is to develop a model to predict the settlement during sheetpile extraction with sufficient accuracy.

As part of this study GeoDelft performed a serie of model tests. The purpose of these tests was to get a better understanding of the phenomena involved in the process of settlement during sheetpile extraction. The test set-up and the first results of analysing the test data will be presented.

## 2. CAUSES OF SETTLEMENT

Different mechanisms can cause settlement of adjacent structures during sheetpile extraction. The most important are:

- densification (plastic volume strain) of the subsoil
- flow of sand to fill the void left by removal of the sheetpile
- flow of liquefied sand
- loss of bearing capacity due to excess pore pressures underneath the building

The cyclic loading of the soil causes the densification and excess pore pressures. This cyclic loading results in

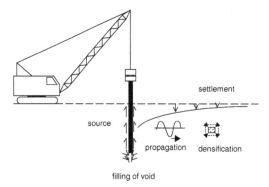

Figure 1. Settlement during sheetpile extraction

a plastic volume strain of the soil. The plastic volume strain depends on many factors e.g.:

– relative density
– shear stress amplitude or shear strain amplitude isotropic stress level
– static deviatoric stress level
– soil structure (aging, effect of preshearing)

In water saturated sand plastic volume strain requires that a volume of water is to be expelled from the soil. In slow loading there is ample time for drainage and the sand behaves drained. During quick loading as in e.g. earthquake loading, time for dissipation is too short and the soil behaves undrained. Excess pore pressures are generated. Afterwards these excess pore pressures dissipate, resulting in a net volume reduction of the soil.

During sheetpile extraction a mixed situation exists. The duration of removing sheetpiles is thus that part of the excess pore pressure can dissipate in the course of this action. The loading frequency is however too high to consider it as drained loading.

During sheetpile extraction also a void is left in the soil by the sheetpile. To fill up this void soil will flow to this void. This will cause some additional settlement in the surrounding.

## 3. EXISTING MODELS

To assess the densification during sheetpile extraction in fact a number of models are to be combined:

– a source model, that describes the vibration at the sheetpile
– a propagation model to assess the response of the soil
– a densification model to assess the plastic volume strain or the excess pore pressure
– possibly a drainage model

Extensive research has been carried out to predict the settlement during earthquake loading, resulting in some simple, semi-empirical models. There are however some significant differences between the situation of earth-

quake loading and sheetpile extraction. The main differences are

– number of cycles (tens of thousands during sheetpile extraction versus 5 to 10 for earthquake loading)
– time of loading (minutes verses tens of seconds)
– stress amplitude

Using this type of models for assessing the densification during sheetpile extraction requires extrapolation of these models. This inevitable leads to large uncertainties. Therefore this type of models cannot be used directly for the densification during sheetpile extraction.

There are some published methods for assessing the densification during sheetpile extraction. All of these methods are (semi) empirical of nature and are based on drained tests. Inevitable many important aspects are not taken into account in these simple models.

Drabkin (Drabkin et al 1996) gives an empirical formula for the settlement as function of different parameters, among them the vibration velocity. The formula is based on a series of cyclic tests. Using values for the different parameters that are outside the tested range results in erroneous results.

Massarch has presented two methods to assess the densification. In (Massarch 1992) the densification is a function of the acceleration and the cone resistance.

In (Massarch 2000) the densification is a function of the amplitude of the shear strain. This model is based on test results by (Youd 1972).

R. Hergarden (Hergarden 2000), (Hergarden, van Tol 2001) has developed a method for assessing the settlement. The densification is a function of the acceleration amplitude. Following the test results of (Barkan 1962) he uses in his model a threshold acceleration below which no densification occurs. He is the only author that that takes into account the settlement due to filling the void left by the sheetpile.

None of these models is capable to predict the excess pore pressure during sheetpile extraction.

Using these models for a specific situation leads to a wide range in predicted settlements. A ratio of 10 between the predicted settlements is not unlikely. With a better understanding of the phenomena involved the risks of unacceptable deformations in the surrounding can be better quantified and thus a better decision for removing the sheetpiles can be made. If it is known beforehand that the sheetpiles cannot be removed, the design can be adjusted such that the sheetpiles can be incorporated in the structure.

## 4. DESCRIPTION MODEL TEST

A serie of tests was conducted in a tank of GeoDelft. The dimensions of the tank are 2*1*1 m. The tank was partly filled with sand (to a depth of approx. 0.6 m) with a uniform relative density of 50%. A plywood sheet is installed in the sand. Tests were conducted with static pulling, vibrating without pulling and a combination of pulling and vibration. The vibrator was connected to a

Figure 2. Model set-up

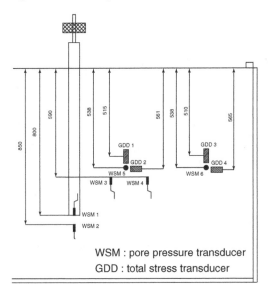

WSM : pore pressure transducer
GDD : total stress transducer

Figure 3. Location of transducers

steel frame, resting on the floor of the hall. The tank itself was placed on this frame as well, separated from it with rubber bearings.

During the tests the following parameters were measured:

– excess pore pressures at the tip of the sheetpile and 5 locations in the soil

– the horizontal and vertical soil stress at two locations
– the acceleration of the sheetpile and of the tank
– the surface of the sand before and after the test
– displacement of coloured dots in the sand
– movement of the soil using a high speed video camera

The aim of the tests was to get an insight in the actual behaviour of the soil during sheetpile extraction.

Of course there are some drawbacks in translating the results of these tests to the prototype situation.

Most obvious are the limited size of the tank and thus the possible influence of the walls of the tank on dissipation of excess pore pressure and reflection of vibrations. Therefore the results are to be considered qualitatively.

## 5. TYPICAL RESULTS MODEL TESTS

The average densification of the sand was assessed by measuring the change in top level of the sand. Figure 4 shows the results of the assessment. On the horizontal axis a parameter $N*v^2$ is used. In this N is the number of vibrations and v the amplitude of the velocity of the sheetpile. This parameter was selected, following the densification model of Sawicki (Sawicki, Swidzinski 1989). In this graph both tests with and without pulling are plotted. The tests without pulling are encircled.

The average trend is that the densification increases with the number of cycles and with an amplitude of the loading. Another trend is that the tests without pulling result in more densification. This can easily be explained as in the tests without pulling the area of the sheetpile in contact with the soil remains the same while for the tests with pulling the area decreases during the test.

There is some scatter around the average trend. With an average volumetric strain of 2 to 3 % in these tests with medium dense sand it is obvious that densification during sheetpile extraction is vibratory pull cannot be denied.

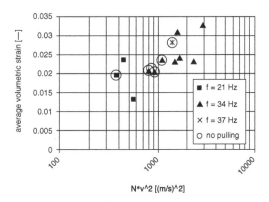

Figure 4. Average densification in performed tests

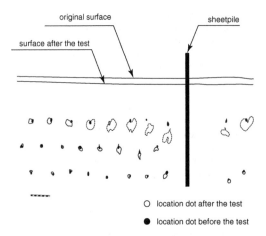

Figure 5. Surface settlement tests DW26C3

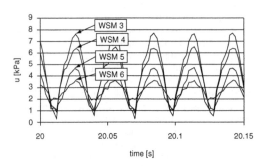

Figure 6. Pore pressure during test DW26C3

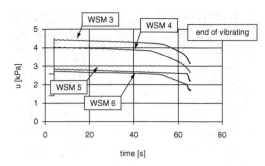

Figure 7. Average pore pressure during test DW26C3

Figure 5 shows the change in surface for test DWD26C3. In this test the thickness of the sheet was 26 mm. The frequency of vibration was 34 Hz and the pulling speed 0.7 cm/s.

The displacements of the dots indicate that the settlement is largest close to the sheetpile and decreases with dept and with distance. The sand surface is quite flat and does not correspond with the displacement profile of the dots. This effect is attributed to flow of the liquefied sand near the surface towards the sheet, thus filling the trough created by the densification.

For the tests with static pulling a settlement trough is observed. The width of this trough was about 30 to 60 cm. This corresponds more or less to an active failure plane, starting at the tip of the sheet.

Figure 6 shows the pore pressure during cycling. Interesting is to examine the development of the excess pore pressure during the test. In figure 7 a moving average of the measured pore pressure is shown for test DW26C3.

Noteworthy is the fast increase in the excess pore pressure. The average pore pressure reaches a value that

is identical to the total vertical stress. This indicates that the sand liquefied during the test.

It is also observed that the average pore pressure decreases during the test. At first this is a slow decrease, corresponding to the settlement of the surface. Towards the end of the test the decrease is faster. It is known that as the soil becomes denser the tendency to densify decreases. Apparently toward the end of the test this tendency has decreased so far that the dissipation of excess pore pressure is in excess of the generation of excess pore pressure. This results in a net decrease in the excess pore pressure.

## 6. CONCLUSIONS

Settlements during sheetpile extraction can be a serious problem. Until now only some relative simple models are available to predict the settlement. Model tests have shown qualitatively some of the mechanisms involved like the generation of excess pore pressure and the flow of liquefied soil. Much work is still to be done to describe the mechanisms involved and incorporate them in a reliable and practical model.

## 7. REFERENCES

Barkan, D. 1962. *Dynamics of bases and foundations.* New York: McGraw-Hill Book Cy Inc.

Clough, G.W. & Chameau, J. 1980. Measured effects of vibratory sheetpile driving. *Journal of Geotechnical Engineering Division*, Vol. 106, No. GT10, October 1980

Drabkin, S. & Lacy, H. & Kim, D.S., 1996. Estimating settlement of sand caused by construction vibration. *Journal of Geotechnical Engineering*, November 1996.

Fujita, K. 1994, Soft ground tunnelling and buried structures. *Proceedings XIII ICSMFE*, New Delhi 1994

Hergarden. R.H. 2000. Grounddeformations during vibratory pull of sheetpiles (in Dutch). *M.Sc. thesis Delft University of Technology, December 2000*

Hergarden, R.H. & Van Tol, A.F. 2001. Settlements during vibratory pull of sheetpiles (in Dutch). *Geotechniek, July 2001*

Lacy, H.S. & Gould, J.P. 1985. Settlement from pile driving in sand. *Proc., ASCE Symposium on Vibration Problems in Geotech. Eng.*, Detroit, Michigan, 1985

Massarch, K.R. 1992. Static and dynamic soil displacements caused by pile driving. *Proceedings 4th Int. Conference Application of Stress-wave Theory to Piles*, The Hague 1992

Massarch, K.R. 2000. Settlements and damage caused by construction-induced vibrations. *Proceedings Intern. Workshop Wave 2000*, Bochum, December 2000

Sawicki, A. & Swidzinski, W. 1989. Mechanics of sandy subsoil subjected to cyclic loading. *Int. Journal for Numerical and Analytical Methods in Geomechanics*, Vol. 13, 1989

Youd, T.L. 1972. Compaction of sand by repeated shear straining. *Journal of the Soil Mechanics and Foundation Division*, Vol. 98, 1972

*Vibratory Pile Driving and Deep Soil Compaction - TRANSVIB2002,*
*Holeyman, VandenBerghe & Charue (eds.), © 2002 Swets & Zeitlinger, Lisse, ISBN 90 5809 521 5*

# Construction process induced vibrations on underground structures

I.Thusyanthan & S.P.G. Madabhushi
*Department of Engineering, University of Cambridge, UK*

ABSTRACT: Vibrations produced on the ground surface by engineering construction processes can damage underground structures. At present there is little knowledge of the level of surface vibrations that could cause damage to underground structures. The relevant British Standards, BS 5228 and BS 7385, have little reference to underground structures. Investigation of the soil-underground structure interaction, under ground borne vibration, must be carried out in a geotechnical centrifuge where prototype stresses and strains are recreated in the model. In this paper we discuss preliminary findings based on 1g experiments on small-scale models. Even though these tests were carried out at low stresses, useful information on the interaction aspects of underground structures was obtained. Experiments were carried out at 1g in a sand model instrumented with an array of miniature accelerometers around two model tunnel inclusions with brass and plastic tunnel linings. Impulsive and harmonic loadings were produced on the sand model surface by a drop hammer mechanism and an electric eccentric-mass motor respectively. The propagation of waves in dry sand and the vibration levels in model tunnels under both the impulse and harmonic surface loading were investigated.

## 1 INTRODUCTION

Engineering construction processes such as piling, blasting, dynamic compaction and demolition produce vibrations to varying degrees. These vibrations are transmitted through the ground as different types of stress wave. When these waves encounter an obstacle such as an underground structure, part of the wave energy is reflected and the rest is transmitted into the structure. The energy transmitted into the structure increases the stress level in the lining of the underground structure. This increase in stress level is usually small compared to the static stresses already present in the structure but since these induced stresses are cyclic in nature they can lead to fatigue cracks and cause damage to underground structures in the long term. Thus it is important to fully understand the propagation of waves through the soil and the transmission of soil vibrations into underground structures.

There are numerous research papers in the literature on the generation and propagation of waves in a half-space; many empirical equations are available to predict the magnitude of vibration away from the source. But there is very little work on the prediction of vibration amplitude in the presence of an underground structure. The relevant empirical equations can be used more effectively with a better understanding of how the presence of an underground structure alters the frequency and amplitude characteristics of the waves. A structure is damaged when the dynamic strains superimposed on the existing strains exceed the tolerance of the structure. The dynamic strain is proportional to the peak particle velocity (ppv). Hence ppv is used to specify the limit on ground vibration that can cause damage to a structure.

Even though the British standards BS 5228 (Part 4:- Code of practice for noise and vibration control applicable to piling operations ), BS 7385 ( Part 2 :- Evaluation and measurement for vibrations in buildings) and the draft Euro code EC3 provide guidance on vibration levels to prevent building damage, there are insufficient case histories to substantiate the guide values. Standards have little or no reference with regard to underground structures. This is mainly due to two reasons. Firstly, underground structures are considered to have a lower degree of risk of damage than the structures above the ground. Secondly, there had not been any recorded major damage to an underground structure due to construction process induced vibrations. This may be because of the present over-conservative limits. However with increased demand on land in major cities, piling operations may need to be undertaken very close to existing underground structures. This demands a thorough understanding of construction process induced vibrations on underground structures.

## 2 APPARATUS AND EXPERIMENTAL TECHNIQUES

### 2.1 *Overview*

Wave propagation in dry sand was studied using impulse and vibrating surface loads on dry sand placed in a 850mm tub. The impulse load was generated by means of a drop hammer mechanism while the vibrating load was generated by an electric eccentric-mass motor.

The 850mm diameter tub was used in all the experiments. Accelerometers, buried at several locations, were used to measure the vibrations in sand. DaisyLab software was used to log the acceleration signals from the accelerometers onto a computer. Experimental apparatus is shown in Figure 1.

Understanding how much of the soil vibration is transferred into an underground structure and how the presence of the structure alters the vibration levels in its vicinity are of real importance. Study in this area was carried out using two model tunnels, made of brass and plastic, placed in the sand tub. All the experiments were carried out at 1g.

### 2.2 *Vibration sources*

Two types of vibration sources were used in the experiment:- impulse load and vibrating load.

### 2.2.1 *Impulse load*

The impulse load was generated by dropping a 5kg mass from a height of 40mm. The drop hammer is controlled by means of a pneumatic switch. When the switch is off, low pressure is created in the vent connected to the hammer; hence the hammer is held in a retracted position. When the switch is on, atmospheric pressure is let into the vent and the hammer falls under gravity onto the base plate (aluminium plate 150mm x 90mm x 15mm), which is placed on the surface of sand.

### 2.2.2 *Vibrating load*

An electric motor with an eccentric rotating mass was used to produce the vibrations on the sand surface. The frequency of the electric motor was 50Hz. The motor was attached to the same base plate, which was used in the impulse load. Two accelerometers were attached to the base plate to record both the horizontal and vertical acceleration-time histories. The base plate experiences a peak particle acceleration of 3g in both horizontal and vertical directions (Fig. 4). This corresponds to component peak particle velocity of 93.7mm/s.

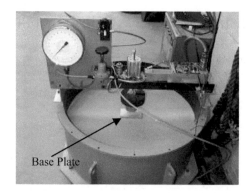

Figure 1. Impulse hammer mechanism on top of sand tub.

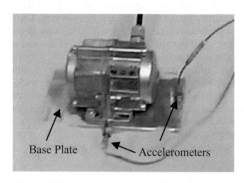

Figure 2. Vibrating motor attached to base plate.

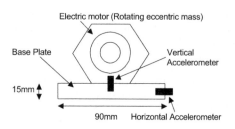

Figure 3. Cross section of vibratory motor and base plate

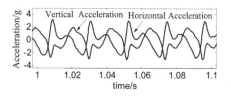

Figure 4. Base plate acceleration versus time.

148

### 2.3 Model preparation and soil type

The quality of the results obtained in the experiment depends directly on the quality of the model. Each time the model was made, the sand was poured into the model with the aid of a hopper. This was to make sure that the sand had the same void ratio and uniformly distributed packing in all the tests. A void ratio of 0.75 was aimed at all the models. LB 100/170 grade E dry sand was used in the experiments. It is uniformly graded sand.

### 2.4 Data acquisition and filtering

Signals from the accelerometers were acquired and recorded using DaisyLab software. A sampling rate of 10kHz per channel was used. The recorded data was post processed using MATLAB before it was used for analysis. Post processing involved eliminating zero-error in the signals and filtering. A butterworth filter was used to eliminate high frequency components above 500Hz.

### 2.5 Model tunnels

Two model tunnels with the same geometry, but different materials, were used in the experiment. The model tunnels, with diameter 54mm and length 320mm, were made of brass and plastic. Brass and plastic were chosen because they have contrasting impedance mismatches with sand (Table 1). Each model had two standard M6 holes at right angles. Accelerometers were securely fastened into these holes to measure the vibrations transmitted into the tunnel from the soil.

Table 1. Impedance of four media.

| Media | Density kg/m$^3$ | $V_p$ * m/s | Impedance kg/m$^2$/s x 10$^3$ |
|---|---|---|---|
| Sand | 1525 | 159 | 242 |
| Plastic | 950 | 996 | 946 |
| Brass | 7500 | 4440 | 33300 |
| Concrete | 2400 | 3316 | 6163 |

* Pressure wave velocity

### 2.6 Experimental setup

12 accelerometers were used in the experiments. 6 accelerometers were placed horizontally at required locations to measure the horizontal accelerations; 6 other accelerometers were placed vertically at exact mirror locations. Three sets of experiments were carried out: Set A, to measure both horizontal and vertical acceleration in dry sand, and Sets B and C to measure the horizontal and vertical acceleration signals in the tunnels and at the tunnel vicinity. 10 tests were carried out in each set (5 impulse load tests and 5 vibratory load tests). Cross sectional views of the models used in experiment sets A, B and C are shown in figure 5. Each test is named using two let-

ters and a test number. First letter represents the set; the second letter represents the type of load (i.e. AV3 - Set A, Vibrating load & Test 3).

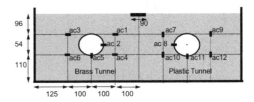

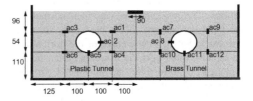

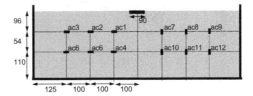

Figure 5. Cross sections of models

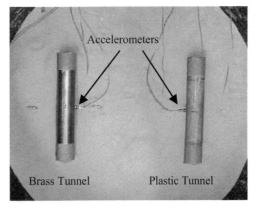

Figure 6. Plan view, preparation of model for set B experiment

149

# 3 RESULTS AND DISCUSIONS

## 3.1 *Results from set A experiment*

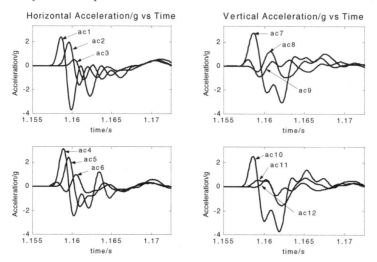

Figure 7. Acceleration signals under impulse load.

### 3.1.1 *Wave fronts from Impulse source*

Resultant peak particle accelerations (ppa) were obtained using horizontal and vertical peak particle accelerations. The resultant ppa was then normalized by the input vertical ppa of the base plate for each experiment in Set A. The average of the normalized ppa at each location, from the five experiments in Set A is plotted below in figure 8. Since sand particles were at rest before the impulse was applied, the direction of peak particle velocity (ppv) would be same as the direction of the ppa. Hence the wave fronts would be at right angles to the ppa vectors. The results agree well with the conventional theory which states that the compression waves propagate radially outwards from the source along a hemispherical wave front (Woods, 1969).

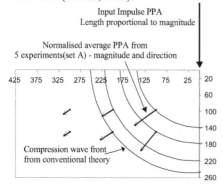

Figure 8. Direction and magnitude of ppa.(dimensions in mm)

### 3.1.2 *Peak particle velocity (ppv)*

Acceleration signals from vibratory load tests were integrated to obtain the velocity-time graphs for all the accelerometer signals. The zero-error in the acceleration signals varies with time. Hence the velocity signal, which results from integrating the acceleration signal, has a non-zero and time-varying mean. This makes the determination of the ppv slightly difficult, but the ppv can be found as half the maximum fluctuation in the velocity signal. Ppv's were obtained from all the signals. Horizontal and vertical ppv's were used to calculate the resultant ppv at the six locations. Resultant ppv's were then normalized using the input vertical ppv of the base plate for each experiment. If we consider spherical wave fronts advancing from the source, then the rate of attenuation of wave energy intensity, due to geometric spreading, would be proportional to $1/s^2$ where s is the slope distance from the source. Peak particle velocity of a wave is proportional to the square root of the energy of the wave. Hence in the absence of material damping, assumable in the present case as the distances between the source and the accelerometers are small, ppv would be expected to diminish as $1/s$. Drawing these lines on the graph shows that the experimental results agree well with the line: - normalized ppv = 9/S (Fig. 9). In this equation, the input energy of the source and the soil condition parameter are both represented by a single value 9.

150

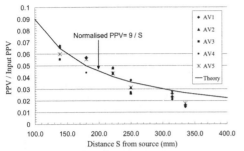

Figure 9. Normalised ppv versus slope distance S from source- results from set A experiments under vibratory load.

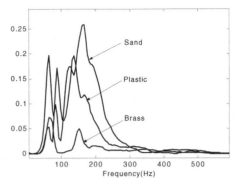

Figure 11. Power spectrum of vertical acceleration signal (ac11) under impulse load.

## 3.2 *Vibration level in model tunnels ( results from set B & C experiments)*

It is clear from figure 10 below that the plastic model tunnel experiences higher peak acceleration in both the horizontal and vertical directions than the brass model tunnel. Specifying the peak acceleration is one way to quantify the difference in the acceleration signals in brass and plastic tunnels, but this does not represent the entire signal. Hence, a better way to quantify the difference in the signals would be to compare the area under the power spectrum of the acceleration signals. The area under the power spectrum represents the energy in the acceleration signal. Thus the ratio of the areas of the power spectrums would represent the energy ratios of the acceleration signals in brass and plastic tunnel.

## 3.3 *Power spectrum analysis*

Figure 11 shows a typical power spectrum of the vertical acceleration signal (ac11) in sand, plastic and brass under impulse surface loading. All three power spectral graphs show two distinct peaks at around 75 Hz and 150 Hz. This corresponds to the natural frequency of the soil in the model, which was calculated to be 75 Hz, and its first harmonic 150 Hz. This shows that the natural frequency of the soil plays a major part in the frequency content of the acceleration experienced by an underground structure.

The average value for the ratio of the power spectral area Brass/Plastic, from ten impulse tests (Table 2), was calculated to be 0.72. Similar power spectral analyses were performed on the acceleration signals from the vibratory load tests.

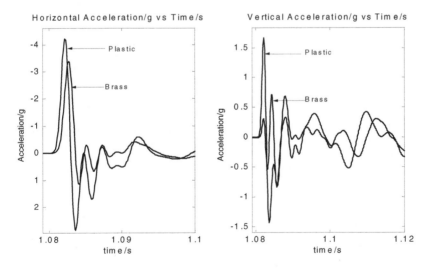

Figure 10. Acceleration signal from Plastic and Brass model tunnels ( test CI2)

151

Table 2. Ratio of power spectral area for vertical acceleration signal under impulse surface load.

| Test | Ratio of power spectral area Brass/Plastic | Test | Ratio of power spectral area Brass/Plastic |
|------|------|------|------|
| BI1 | 0.60 | CI1 | 0.29* |
| BI2 | 0.73 | CI2 | 1.00* |
| BI3 | 0.72 | CI3 | 0.71 |
| BI4 | 0.69 | CI4 | 0.81 |
| BI5 | 0.64 | CI5 | 0.86 |

* These results were excluded in the average, as they do not follow the general trend.

The average value for the ratio of the power spectral area Brass/Plastic, from ten vibratory tests (Table 3), was calculated to be 0.73. Figure 12 shows a typical power spectrum of the vertical acceleration signal in plastic and brass under vibratory surface loading. The power spectrum of the signal in sand is not shown, as it is very similar to that of plastic but with higher magnitude. It is evident from the power spectrum that the vertical acceleration signal of the brass tunnel has the most energy near 100Hz while that of the plastic tunnel has the most energy near 150Hz. This trend is also exhibited in figure 11. This suggests that brass and plastic transmit energy at different frequencies in addition to the fact that they transmit different quantities of energy.

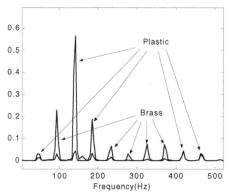

Figure 12. Power spectrum of vertical acceleration signal (ac11) under vibratory load.

Table 3. Ratio of power spectral area for vertical acceleration signal under vibratory surface load.

| Test | Ratio of power spectral area Brass/Plastic | Test | Ratio of power spectral area Brass/Plastic |
|------|------|------|------|
| BV1 | 0.76 | CV1 | 0.79 |
| BV2 | 0.68 | CV2 | 0.82 |
| BV3 | 0.62 | CV3 | 0.75 |
| BV4 | 0.67 | CV4 | 0.73 |
| BV5 | 0.75 | CV5 | 0.75 |

The impedance mismatch between the sand and the tunnel determines the amount of wave energy transmitted into the tunnel. The impedance mismatch between sand and brass is greater than that between sand and plastic. Hence more energy will be transmitted into plastic tunnel. Results from the above power spectral analysis suggest that energy transmitted into a brass tunnel is 72.5% of that transmitted into a plastic tunnel.

An alternative method to quantify the ratio of energy transferred into the model tunnels is to use peak particle velocity. Acceleration signals were integrated to obtain the velocity-time graphs. Peak particle velocities were obtained from all the velocity–time graphs. Table 4 summarises the ratio of ppv in brass to plastic in all ten tests. The average ratio of ppv in brass to plastic was calculated to be 0.82.

Table 4. . Peak particle velocity (mm/s) of model tunnels in vertical direction.

| Test | Brass ppv mm/s | Plastic ppv mm/s | Brass ppv / Plastic ppv |
|------|------|------|------|
| BV1 | 1.40 | 1.80 | 0.78 |
| BV2 | 1.50 | 1.75 | 0.86 |
| BV3 | 1.30 | 1.50 | 0.87 |
| BV4 | 1.35 | 1.60 | 0.84 |
| BV5 | 1.25 | 1.40 | 0.89 |
| CV1 | 1.25 | 1.50 | 0.83 |
| CV2 | 1.22 | 1.50 | 0.83 |
| CV3 | 1.18 | 1.30 | 0.90 |
| CV4 | 1.10 | 1.40 | 0.79 |
| CV5 | 0.90 | 1.40 | 0.64 |

### 3.4 Relationship between energy transferred and impedance mismatch

Energy transferred into the model tunnel is proportional to the square of the peak particle velocity in the model tunnel. Hence the ratio of energy transferred into the model tunnels can be calculated using the ppv ratio of brass to plastic.

Energy transferred into Brass model tunnel

---

Energy transferred into Plastic model tunnel

$$= 0.82^2 \quad = 67\%$$

We can try to correlate the impedance mismatch ratio to the square of the ppv ratio. Table 1 shows the impedances of four media. Let the impedance of sand, plastic and brass be $I_s$, $I_p$ and $I_b$ respectively. The following relationship is proposed:

$$\left[ \frac{Brass's\ ppv}{T's\ ppv} \right]^2 = \left[ \frac{I_T - I_s}{I_b - I_s} \right]^n \tag{1}$$

where n is a constant; T is a model tunnel whose impedance is between that of brass and plastic. Figure

13 shows the above relationship lines for n=0.05, 0.1 and 0.15. Note that all three lines pass through the boundary condition (i.e. the ratio of energy transferred is one when the ratio of impedance mismatch is one). It can be seen from figure 13 that the line corresponding to n=0.1 agrees well with the experimental point for plastic (ratio of brass to plastic ppv squared = 0.67).

Figure 13 can be used to predict energy transferred into a material T at shallow depths (at low soil stresses). Thus we can predict that the ratio of energy transferred into a brass tunnel to a concrete tunnel, at shallow depths, is 0.86.

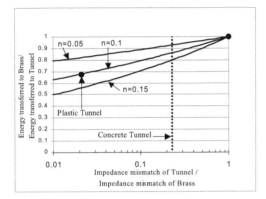

Figure 13. Ratio of Energy transferred vs ratio of impedance mismatch

## 4 CONCLUSION

The tunnel lining has an important role to play in determining the amount of energy absorbed from ground borne vibrations. The vibration amplitude transferred into the plastic tunnel was shown to be higher than that transferred into the brass tunnel under impulse and harmonic loads. Under harmonic loading, the plastic model tunnel appeared to absorb energy at higher frequencies (150Hz-200Hz) relative to brass model tunnel (100Hz). Experimental results show that the ratio of peak particle velocity (vertical) in the brass tunnel to the plastic tunnel is 0.82. Hence the ratio of energy transferred into the brass tunnel to the plastic tunnel, which is the square of ppv ratio, is 0.67.

It is also worth remembering that the damage to an underground structure is not only dependent on the amount of energy transferred into the structure but also on the frequencies at which the energy is transferred.

The following relationship is proposed for the impedance mismatch ratio and the square of the ppv ratio (i.e ratio of the energy transferred).

$$\left[ \frac{Brass's\ ppv}{T's\ ppv} \right]^2 = \left[ \frac{I_T - I_s}{I_b - I_s} \right]^{0.1} \qquad (2)$$

The above conclusions can form the basis on which more research can be carried out to expand the knowledge in this field. The energy transfer into various tunnel linings such as pre-cast concrete, shotcrete and steel can also be investigated. Thorough study in this field will enable us to understand and improve on the vibration limits set out by the present British Standards.

## REFERENCES

Chao, C.-C.(1960) *Dynamical response of an elastic half-space to tangential surface loadings.* J. appl. Mech. 27, 559-67.
Head J.M and Jardine F.M.(1992) *Ground borne vibrations arising from piling.* CIRIA-Technical Note 142.
Hiller D.M. and Bowers K.H.(1997). *Groundborne vibrations from mechanized tunnelling works.* Proceedings of Tunnelling 97, Institution of Mining and Metallurgy, London, 1997, 721-735.
Hope V.S and Hiller D.M.(1998) *The Prediction of groundborne vibration from percussive piling.* Canadian Geotechnical Journal, Vol-37,Pg 700-711.
Hudson. J.A (1980) *The excitation and propagation of elastic waves.* Cambridge University Press.
Karl F.Graff (1975). *Wave motion in Elastic solids* . Oxford University Press.Pg 395-410.
Madabhushi S.P.G and Steedman R.S.(1989). *Wave transmission at a multi-media interface.* Technical report, CUED/D-SOILS/TR227, Cambridge University.
Miller, G.F. and Pursey, H.(1954) The Field and the radiation impedance of mechanical radiators on the free surface of a semi-infinite isotropic solid. Proc.R.Soc A223, 521-41.
Thusyanthan, I. (2001) *Construction process induced vibrations on underground structures.* MEng Project Report, Cambridge University.
Woods R.D, Richard F.E, Hall J.R.(1969).*Vibrations of soils and foundation.* Prentice-Hall.

*Soil investigation and bearing capacity*

# Development of a Vibro-Penetration Test (VPT) For In-situ Investigation of Cohesionless Soils

Roberto O. Cudmani & Gerhard Huber
*Institute of Soil and Mechanics, University of Karlsruhe*

A new dynamic penetration testing for in-situ investigation of cohesionless soils based on the principles of vibratory driving has been developed. Basically, the so-called vibro-penetration test (VPT) differs from blow-type dynamic penetration testing methods in the driving force and the mode of penetration. During the VPT a penetrometer is driven by a harmonic excitation force whereby the driving velocity and acceleration at the tip are continuously recorded. The vibro-penetration resistance is evaluated from the recorded data on the basis of a suitable mechanical model. The result of a VPT is given as a diagram showing the penetration resistance verses depth. This is defined as the number of cycles Nz10 needed to drive the penetrometer tip 0.10 m downwards with a reference driving energy. It is concluded that the quality of the results of VPT is comparable with that come penetration test (CPT) and better than other dynamic penetration tests, e.g. standard penetration tests (SPT) and dynamic probing test (DPT), if the soil response at the tip is recorded and correctly interpreted. The advantages of VPT as against CPT are a smaller static load required for driving, shorter execution time and its suitability for in-situ investigations of medium dense to dense cohesionless soils at large depths.

## 1 INTRODUCTION

The mechanical behaviour of simple cohesionless grain skeletons is determined by granulometric properties, such as grain shape, grain mineral and grain size distribution, as well as by the variables defining the state of the material, in particular density and stress. Granulometric properties can be investigated in the laboratory using disturbed soil samples, whereas the state of the material can be only evaluated either from undisturbed samples or indirectly by using in-situ testing procedures. Penetration tests are among the available in-situ testing techniques mostly used in geotechnical practice, since they are fast and economical and provide a rather reliable basis for the empirical or analytical estimation of the in-situ density. Penetration testing techniques can be divided into dynamic and static. Dynamic penetration tests are easy to carry out and require relatively simple equipment permitting frequent tests. However, as is well known, the blow count as a measure of penetration resistance is strongly affected by variations of the energy delivered by the hammer, as well as by other factors like rod friction and type of hammer. On the other hand, static penetration tests have the advantages of simplicity, continuous data recording, reproducibility of results and suitability of the test data for rational analysis. However, although static methods are qualitatively superior to dynamical ones, the use of the latter can be inevitable in case the soil cannot be penetrated statically,

such as medium dense or dense cohesionless soil deposits, especially at large depths.

In the vibro-penetration test (VPT), the dynamic penetration force is generated by vibro-driver rather than by blows. Thus, the probe is driven into the soil by a vertical harmonic excitation, which results from the combination of a dynamic force generated by a vibrator and the weight of a vibration isolated bias mass.

As in conventional dynamic penetration tests the driving resistance in the VPT is in a first approximation, inversely proportional to the driving velocity. A change of the machine parameters of the vibro-driver, equivalent to a change of the blow energy in blow-based testing methods, and the increase of the rod length and friction resistance with increasing testing depth, may alter the energy available to displace the soil at the tip. Such alterations, which are neither related to variations of soil properties nor to changes of the soil states over depth, may cause ambiguous results if one uses penetration rate as a measure of penetration resistance without taking into account the energy released at the tip for driving.

Two conditions must be necessarily fulfilled to overcome this difficulty. Firstly, one needs to measure the penetration rate of the tip and the energy spent to drive it into the soil. Secondly, a reliable relationship between the penetration rate and the amount of energy spent for driving the probe must be determined. The last condition provides a systematic way to compare penetration rates obtained for different energies. The main advantage of VPT with respect to SPT and DPT is that both conditions can be fulfilled quite and much more easily.

The VPT-prototype used in our experimental investigations will be shown in Section 2. Two qualitatively different types of vibro-penetration, called *cavitation* and *no-cavitation* modes (CUDMANI (1), CUDMANI et al. (2)), the prevalence of which depends on the setting of machine parameters and on the properties and the soil state, will be described in Section 3. Understanding cavitation and no-cavitation modes of penetration from a soil mechanics point of view is the prerequisite to develop the method presented in Section 4 for the evaluation of the vibro-penetration resistance. Based on the side friction near the tip, a quantity for the investigation of soil stratigraphy will be defined in Section 5. Thereafter the reliability of the VPT will be verified by calibration chamber tests in the laboratory (Section 6).

## 2 PROTOTYPE OF A VPT-DEVICE

The VPT-prototype consists of a vibratory driver, a guiding frame, driving rods and an instrumented tip as illustrated schematically in Figure 1. The vibratory driver consists of a vibrator, bias mass, isolation springs and a hydraulic clamp for connecting the vibrator and the drive rods. The vibrator produces vertical forces by means of counterrotating eccentric masses, the rotation of which is induced by a hydraulic motor connected to a 30 kW hydraulic supply. The machine parameters controlling the magnitude of the dynamical force and the vibration amplitude amplitude are the static moment of the eccentric masses with respect to its rotation axes ($S_v$), its rotation frequency ($f$) and the vibrating mass ($m$). The centrifugal force generated by the conterrotating masses is given by:

$$F_0 = 4\pi^2 f^2 S_v \sin (2\pi ft) \qquad (1)$$

The displacement amplitude of the vibrator mass ($m$) in the air is given by:

$$u_0 = \frac{S_v}{m} \qquad (2)$$

A vibro-driver with maximum force amplitude of 80 kN and a mass of about 150 kg was designed and built. Operating frequencies and displacement amplitudes of the vibrator can be stepwise varied between 25 and 100 Hz and 1.3 and 5.3 mm. For comparison: low frequency vibrators (up to 100 Hz) currently used for the vibro-driving of piles and sheet piles have maximum force amplitudes varying between 400 until 5000 kN. The isolation springs were designed for a maximum bias weight of 1,5 kN. The latter can be easily varied by means of the counterweight system shown in Figure 1.

The guiding frame, which is necessary to guide both the vibro-driver and the rods in the vertical direction, consists of a platform and two lateral guiding profiles with a total weight of 10 kN. The platform is provided with four adjustable legs which permit the levelling of the guiding frame.

The driving rod consists of 3,20 m long stainless steel tube segments having an outer diameter of 36 mm and a wall thickness of 20 mm. The rod segments weight about 20 kg. They are provided with vibration resistant screw joints which prevent a loosening of the segment connections during driving. The geometry and instrumentation of the VPT-penetrometer tip are similar to those of the reference electrical CPT-penetrometer tip (DE BEER et al., (5)). The penetrometer tip has a diameter of 36 mm and ends in a cone with an apex angle of 60°. The friction sleeve, the area of the penetrometer tip upon which the local side friction resistance is measured, is located immediately above the cone and has a length of 133 mm with an area of 15042 mm². The soil resistances at the tip and at the friction sleeve are measured with separated load cells. In order to measure the local motion of the penetrometer tip an accelerometer is placed at the tip above the load cells. In addition, an inductive displacement transducer attached to the bias mass is used to measure the penetration of the penetrometer. In order to monitor the working frequency of the vibrator, the position of one of the eccentric masses is continuously measured using a magnetic gauge.

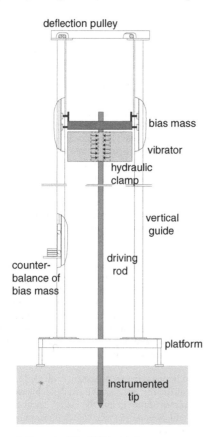

deflection pulley

bias mass

vibrator

hydraulic clamp

vertical guide

counter-balance of bias mass

driving rod

platform

instrumented tip

Figure 1. Layout of the VPT-prototype.

Figure 2. VPT-prototype.

The five electrical signals are amplified and filtered before being digitized and acquired. The data acquisition is carried out using a 8 channel data acquisition card installed on a PC (Pentium I, 133 MHz, 32 MB RAM) with an acquisition rate of 4000 samples/s.

## 3 PENETRATION MODES DURING VIBRATORY DRIVING

Experimental results shows that the of penetration during vibratory driving is strongly influenced by the soil resistance at the tip. Therefore, the development of a method to evaluate the vibro-penetration resistance based on the rate of penetration requires the understanding and the modelling of the mechanical processes at the tip.

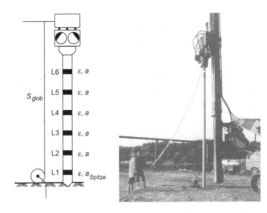

Figure 3. Full-scale investigation of vibratory driving: instrumented pile.

First, we will consider here the experimental results of full-scale vibratory driving tests with an instrumented pile (HUBER (8)). The steel pile was a tube with a diameter of 0.160 m and a length of 7.25 m. It was instrumented in 6 levels with strain gauges and accelerometers for measuring the vertical force and the vertical acceleration (Figure 3). The global displacement was measured continuously with an electrical rope drum gauge.

Figure 4 shows two typical soil responses at the tip (force-displacement-response) for five driving loops in a depth of 5 m which were obtained in two driving tests by changing the of the setting of the machine parameters. The tests were carried out in the same test site (near Hagenbach, Germany) in two closely located testing positions. Both the ground of the test site, which consists of layers of sand and silty sand, and the initial state were relatively homogeneous so that similar soil and state conditions for both driving tests can be assumed. The example shows clearly that the machine parameters can influence the evolution of the tip resistance not only quantitatively but also qualitatively.

Let us consider one of the force-displacement loops in Figure 4a. With the help of Figure 5a, which is an idealization of the measured response, four well phases can be identified within a loop. We begin at the maximum force in point 1 (phase I). The pile moves and the tip remains in contact with the soil. After a relatively small upward displacement (point 2) the contact between the soil and the pile gets lost, since the piles moves faster at this point than the soil. The pile the underlying soil uncouples and a cavity forms underneath the tip (phase II). From point 3 (motion reversal) the pile tip moves downward without contact with the soil until point 4 (phase III), which does not coincide with point 2. Between point 2 and 4 the force-displacement curve does not give any information about the soil response. After the contact is restored the soil resistance is mobilized again (phase IV). At the end of a loop (point 1') the

159

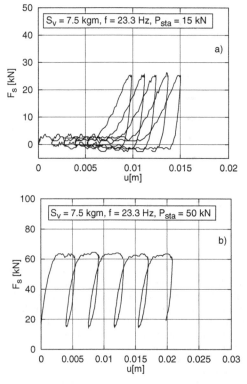

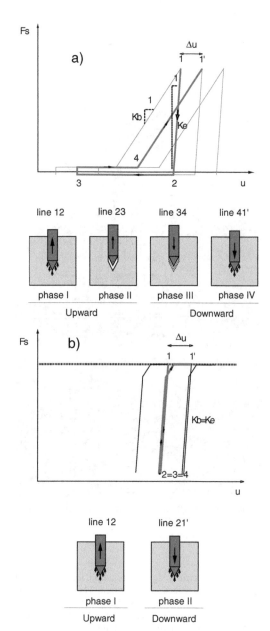

Figure 4. force-displacement loops measured at the tip of the instrumented pile: a) cavitation b) no-cavitation modes of penetration.

pile has been diven the amount $\Delta u$. The relatively large reversibility of deformation observed between points 2 and 4 is mainly due to the vertical stretching of the neighbouring soil elements during unloading. The vertical stretching happens because the reduction of vertical stresses is stronger than the reduction of horizontal stresses. The soil is sheared nearly as in a triaxial extension test. It must be pointed out that contrary to the common assumption the pile is driven *without* achieving a limit resistance. Due to the separation of pile tip and soil during unloading this penetration type is called *cavitation* mode. The response curve for the cavitation mode can be determined by three quantities:

- the inclination $K_b$ of the downward displacement line 4–1′
- the inclination $K_e$ of the upward displacement line 1–2
- the penetration per loop $\Delta u$

For computing purposes, it is useful to express the penetration $\Delta u$ as a function of the plastic deformation ratio $\beta_p$ which is defined as:

$$\beta_p = \frac{\Delta u}{u_4 - u_{1'}} \qquad (3)$$

Figure 5. Idealization of mechanical soil response at the pile tip: a) cavitation b) no-cavitation mode of penetration.

$\beta_p$ varies between 0 and 1. $\beta$=0 corresponds to $\Delta u$=0, i.e. points 1 and 1′ coincide and there is no penetration. If points 2 and 4 coincide $\beta_p$=1.

Let us consider one of the force-displacement loops in Figure 4b. In this case two motion phases can be recognized within a loop (see also idealisation in

160

Figure 5b). Different from the cavitation mode the contact between the pile tip and the underlying soil is not lost during the upward motion phase (phase I). Thus, the soil resistance is mobilized immediately after the reversal of the pile motion (points 2, 3 and 4 coincide). During the downward motion phase, the soil resistance increases almost as fast as it decreases during unloading until a limit value is achieved. After this point the downward motion continues at constant soil resistance until the next reversal of motion at point 1' takes place. Since no separation between pile and soil takes place during driving we call this penetration type *no-cavitation* mode. In order to characterize this driving mode, the following quantities are used:

- the inclination $K_b \approx K_e$ of the downward and upward displacement lines
- the limit resistance $F_{s,max}$
- the penetration per loop $\Delta u$

More details of this theory are outlined by CUDMANI et al. (4) in this Volume. Two force-displacements loops at the tip of VPT carried in the calibration chamber (see Section 6) are shown in Figure 6. In both tests the same setting of machine parameters was used, but the initial state of the soil was different (relative density $I_d = 0.27$ and 0.51; mean pressure $p_0 = 100$ and 200 k Pa). These diagrams show that not only a change of the parameter setting determines the occurrence of cavitation or no-cavitation modes of penetration, but also the initial state of the soil.

A soil mechanical justification for the occurrence of cavitation and no-cavitation modes was found to be the strong dependence of granular materials behavior on the most recent deformation history for alternating loading (CUDMANI (3)). In the cavitation mode driving the stretching of the soil during the upward motion leads the soil near the pile tip to a limit state, and there the memory of the material on the previous deformation history is almost totally swept out (GUDEHUS et al. (7)). The soil response is determined only by the density and the state of stresses just before the reversal of motion takes place. Whereas in the no-cavitation mode the soil is not completely unloaded during the upward motion phase and the shear deformation is not large enough to erase the material memory.

## 4 EVALUATION OF THE VIBRO-PENETRATION RESISTANCE

The vibro-penetration resistance is defined by the number of cycles $N_{z10}^*$ of the excitation force which are necessary to drive the probe 0.10 m into the soil. $N_{z10}^*$ is the ratio of the reference penetration $\Delta r = 0.10$ m to the penetration per loop $\Delta u = v_{glob}/f$:

$$N_{z10}^* = \frac{\Delta r}{\Delta u} = \frac{0.10\,f}{v_{glob}} \qquad (4)$$

The vibro-penetration resistance given by equation 4 is ambiguous since the global driving velocity ($v_{glob}$) depends not only on the properties and state of the soil, but also on other quantities like the setting of the machine parameters, the friction forces acting on the driving rod and the damping of the guide-vibrator-rod system. As it was pointed out in the introduction, the correct evaluation of the vibro-penetration resistance requires to take into account the fraction of the energy released at the tip responsible for driving the penetrometer.

The relationship between the energy delivered by a blow impact and the penetration resistance for different cohesionless soils was experimentally investigated for SPT and DPT (PALACIOS & SCHMERTMANN (11), ROBERTSON et al. (10), KRAMER (9)). It was found that for given granulometric properties and soil state the blow count $N^*$ ($N^* \sim 1/\Delta u$) is inversely proportional to the energy released at the tip, i.e.

$$N^*\,W_{blow} = const. \qquad (5)$$

In this case $W_{blow}$, which is the blow energy $E = Gh$ ($G$: weight of the falling mass, $h$: height of fall) can be determined by knowledge of the impact efficiency factor $\eta$:

$$W_{blow} = \eta G\,h \qquad (6)$$

Using equation 5 blow counts obtained with different blow energies can be compared with each other by

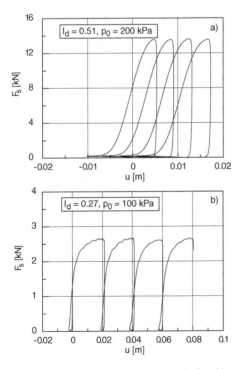

Figure 6. VPT in the laboratory: a) cavitation b) no-cavitation modes of penetration.

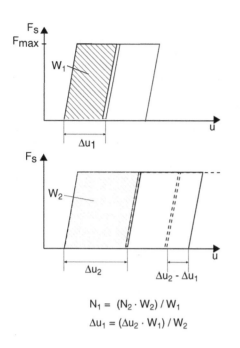

$$N_1 = (N_2 \cdot W_2) / W_1$$

$$\Delta u_1 = (\Delta u_2 \cdot W_1) / W_2$$

Figure 7. Relationship between penetration resistance N and released energy $W$ for the no-cavitation mode.

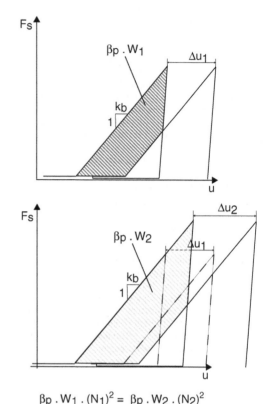

$$\beta p \cdot W_1 \cdot (N_1)^2 = \beta p \cdot W_2 \cdot (N_2)^2$$

$$\beta p \cdot W_1 \cdot (1 / \Delta u_1)^2 = \beta p \cdot W_2 \cdot (1 / \Delta u_2)^2$$

Figure 8. Relationship between penetration resistance N and released energy $W$ for the cavitation mode.

taking one of the employed blow energies as a reference energy $E_r$,

$$N = N^* = \frac{W_{blow}}{E_r} \qquad (7)$$

Equation 7 indicates that in terms of the vibratory driving modes described in Section 3 impact driving can be classified as a no-cavitation driving process. As shown in Figure 7 a change of energy spent at the tip per cycle $W_z$, which is determined by the area enclosed in one force-displacement loop, induces a proportional change of $\Delta u$ ($\Delta u \sim N_{z10}^*$) as predicted by equation 7.

In a similar manner a relationship between $N_{z10}^*$ and the mechanical work per cycle $W_z$ for the cavitation mode can be found using the mechanical model proposed by DIERSSEN (6). The model describes the dependence of the tip resistance on displacements for the cavitation mode as a function of the quantities $K_b$, $K_e$ and $B_p$. Assuming that for given geometry and soil properties these quantities depend only on the soil state we obtain analog to equation 5 the following relationship between $N_{z10}^*$ and $W_z$ for the cavitation mode:

$$N_{z10}^* \, \beta_p \, W_z = const. \qquad (8)$$

The derivation and the geometrical interpretation of equation 8 is illustrated in Figure 8. An experimental investigation shows that the assumption made above is not strictly valid for vibration frequencies and static moments differing more than 20% from each other (CUDMANI (3)). For this reason these two machine parameters were kept nearly unchanged in all VPT ($f \approx$ 30Hz, $S_v = 0.81$ kgm). The relationship between the vibro-penetration resistances $N_{z10}^*$ and $N_{z10}$ obtained for $W_z$ and a reference energy $E_e$, respectively is:

$$N_{z10} = N_{z10}^* \sqrt{\frac{\beta_p W_z}{E_r}} \qquad (9)$$

The maximum kinetic energy of the vibrator for a frequency $f = 25$ Hz, a mass $m = 100$ kg and a static moment $S_v = 0.81$ kgm, $E_r \approx 54$ Nm, was adopted as reference energy for the evaluation of VPT.

In order to evaluate $N_{z10}$ verses depth the current penetration rate $v_{glob}$, the current frequency $f$ as well as the mechanical work of the tip force $W_z$ were determined for depth intervals. For the calculation of $W_z$ four periods of the tip force $F_s(t)$ and the tip velocity $v_s(t)$ were considered in the chosen evaluation depths. The velocity of the tip was obtained from the sum of the global driving velocity and the local velocity which results from the integration of the tip acceleration $a(t)$. The mechanical work $W_z$ results from the integration of the current power $P(t)$ from $t_0$ to $t_0 + 4T$ ($T = 1/f$):

$$W_z = \frac{1}{4}\int_{t_0}^{t_0+4T} P(t)dt = \frac{1}{4}\int_{t_0}^{t_0+4T} F_s(t)v_s(t)dt \qquad (10)$$

Theoretically, equation 8 is valid for any driving velocity provided that $v_{glob} > 0$ (if the probe sticks $N_{z10}^*$ goes to infinity and the product $\beta_p W_z$ approachs zero). Practically, there is a minimal driving velocity, which depends on the test equipment, below which the calculated vibro-penetration resistance loses reliability. For the VPT-prototype, the minimal reliable driving velocity is 0.005 m/s.

Equations 8 and 9 apply only to the cavitation mode. For the evaluation of VPT in connection with the no-cavitation mode equation 7 must be used. However, vibro-penetration resistances obtained on the basis of different penetration modes are not convertible. Therefore, the setting of the machine parameters should be appropriately chosen in order to force the occurrence of only one driving mode. Since the generation of the no-cavitation mode requires usually larger quasi-static forces, it was easier with our VPT-prototype to generate the cavitation vibratory mode.

## 5 FRICTION RATIO FROM THE VPT

Figure 9a shows four loops of the shear stress-displacement response obtained at the shaft of the instrumented pile between the levels 1 and 2 (see Figure 4). Figure 9b illustrates the same response for the friction sleeve of the instrumented tip. It can be seen that the measured response is similar in both cases. After each reversal the shear stresses increase quite rapidly up to a maximum value $\tau_{max}$, which is almost independent of the direction of motion. Using this property we can define a quantity for the investigation of the soil stratigraphy in analogy to the friction ratio in the CPT. For the cavitation mode we propose the quantity:

$$f_R = \frac{\tau_{max}A_s}{K_b\beta_p d_c} \qquad (11)$$

Wherein are $A_s = 15042$ mm$^2$ and $d_c = 36$ mm. For the no-cavitation mode the friction ratio $f_R$ is defined as for CPT,

$$f_R = \frac{\tau_{max}A_c}{F_{s,max}A_s} \qquad (12)$$

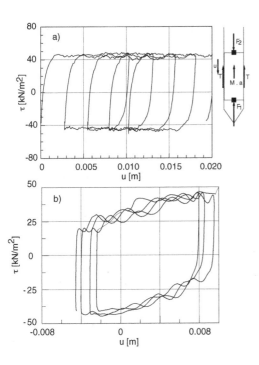

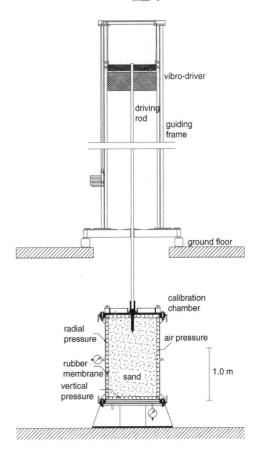

Figure 9. Friction resistance during vibratory driving: a) instrumented pile, b) friction sleeve of the instrumented penetrometer tip.

Figure 10. Layout of the VPT-prototype and the calibration chamber for the laboratory investigation.

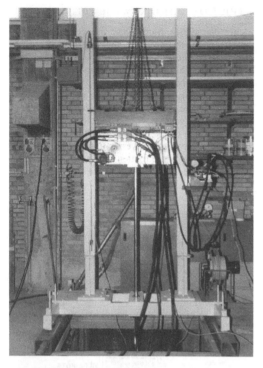

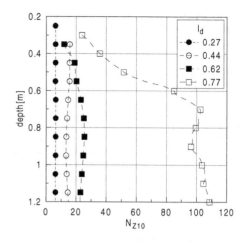

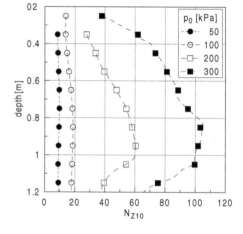

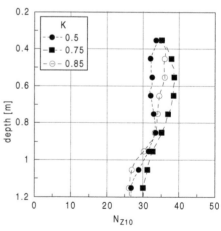

Figure 11. VPT-prototype and calibration chamber for the laboratory investigation.

Figure 12. $N_{z10}$ over depth for different $I_d$ and $p_0 \approx$ 100 kPa (top), for different $p_0$ and $I_d \approx 0.5$ (middle) and for different $K$ and $I_d \approx 0.5$ and $p_0$=200 kPa (bottom).

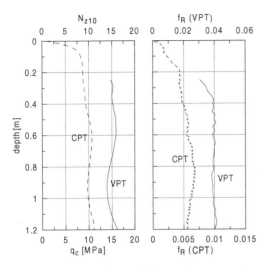

Figure 13. Comparison of the static penetration resistance $q_c$ and Vibro-penetration resistance $N_{z10}$ (left) and friction ratio $f_R$ from CPT and VPT (right).

## 6 EXPERIMENTAL VERIFICATION OF THE VPT IN THE LABORATORY

In order to determine a relationship between $N_{z10}$ and the state variables VPT were carried out in a calibration chamber for different initial soil states. The calibration chamber is a cylindrical steel container with $D = 0.94$ m and $H = 1.50$ m (Figure 10, 11). In the tests a dry uniform medium quartz sand ($d_{50}$=0.4 mm; $e_{max}$=0.84; $e_{min}$=0.53) was used, which was placed by the pluviation method. Lateral and vertical pressures were applied independently to the soil column to simulate in-situ stress conditions, both instropic and anistropic. A detailed description of the used equipment, the sand placement method and the testing program can be found elsewhere (CUDMANI (3)). Typical results of VPT in the laboratory are presented in Figure 12.

The diagrams show the variation of the vibro-penetration resistance as a function of the relative density $I_d$, the far-field mean pressure $p_0$ and the initial stress ratio $K$. $N_{z10}$ shows qualitatively the same dependence on the soil state as $q_c$ or $N_{SPT}$, i.e an increase of $I_d$ and $p_0$ leads to an increase of the vibro-penetration resistance. Figure 13 presents a comparison of VPT and CPT results for the same initial state. Both the vibro-penetration and the static penetration resistance reveal similar variations over depth.

## 7 CONCLUSIONS

The soil resistance is dominant for the penetration progress during vibratory driving of piles. The tip motion is characterized by two different penetration modes whose prevalence depends on the setting of the parameters of the vibro-driver as well as the soil state. The cavitation mode takes place when the contact between the tip and the soil is momentarily lost during the upward tip motion. In contrast, in the no-cavitation mode the tip and the soil does not loose contact during driving. Understanding the cavitation and no-cavitation modes was the prerequisite for the development of the vibro-penetration testing method. Advantages of the VPT over CPT are the short duration, the considerably smaller static load required for penetration (about 20 times smaller) as well as the suitability for medium dense to dense cohesionless soils at larger depths. VPT differs from blow-based-penetration tests in the mode of penetration since, as experimental results of SPT and DPT in the literature showed, in this case the probe penetrates in the no-cavitation mode. The vibro-penetration resistance only measured as the penetration rate can be ambiguous. Changes of the machine parameters induce changes of penetration rate which are not related to variations of the soil state. For this reason a suitable mechanical model was proposed to analyse the vibro-penetration records, taking into account the penetration rate, the released energy at the tip as well as the penetration mode on the vibro-penetration resistance. In the laboratory and in the field VPT-results presented a good reproducibility and correlated well with the results of CPT.

## 8 ACKNOWLEDGEMENT

The Development of the VPT was part of the research project "Reconstruction and Stabilisation of Dumps and Dump Slopes Endangered by Settlement Flow" supported by the LMBV mbH (Lusatian and Central German Mines Administration Company) and the German Federal Ministry of Education and Research (BMBF).

## REFERENCES

1. R. Cudmani: A Soil Mechanical Model for Modelling Vibratory Driving in Cohesionless Soils. *Workship Vibratory Driving*, Lehrstuhl für Bodenmechanik und Grundbau am Institut fü Boden- und Felsmechanik der Universität Karlsruhe, 1997 (in German).
2. R. Cudmani, G. Huber und G. Gudehus: Cyclic and Dynamic Penetration of Non-cohesionless Soils. In: *Boden unter fast zyklischer Belastung: Erfahrungen und Forschungsergebnisse*, Schriftenreihe des Institutes für Grundbau und Boden-mechanik der Ruhr-Universität Bochum, Th. Triantafyllidis (ed.), 2000 (in German).
3. R. Cudmani: Static, Alternating and Dynamic Penetration of Cohesionless Soils. Veröffentlichung des Institutes für Boden-mechanik und Felsmechanik der Universität Karlsruhe, Nr. 152, 2001 (in German).

4. R. Cudmani, G. Huber, G. Gudehus: A mechanical Model for the Investigation of the Vibro-Drivability of Piles in Cohesionless Soils. *Proceeding* of the International Conference on Vibratory Driving and Deep Compaction, 2002 (in this volume).

5. E.E. De Beer, E. Goelen, W.J. Heynen, K. Joustra: Cone Penetration Test (CPT): International reference procedure. *Proceedings* of the International Symposium on Penetration Testing. De Ruiter (ed.), Orlando, 1988.

6. G. Dierssen: Ein Bodenmechanisches Modell zur Beschreibung des Vibrationsrammens in Körnigen Böden. *Veröffentlichungen des Institutes für Bodenmechanik und Felsmechanik der Universität Fridericiana in Karlsruhe*, Heft 133, 1994 (in German).

7. G. Gudehus, M. Goldscheider und H. Winter: Mechanical Properties of Sand and Numerical Integration Methods, Some Sources of Errors and Bounds of Accuracy. In: *Finite Elements in Geomechanics*, Gudehus(ed.), John Wiley, New York, 1977.

8. G. Huber: Vibratory driving. Full-scale Field Tests. *Workshop Vibratory driving*, Lehrstuhl für Bodenmechanik und Grundbau am Institut fü Boden- und Felsmechanik der Universität Karlsruhe, 1997 (in German).

9. H.-J. Krämer: Geräteseitige Einflußparameter bei Ramm- und Drucksondierungen – und Ihre Auswirkungen auf den Eindringwiderstand. *Veröffentlichungen des Instituts für Maschinenwesen im Baubetrieb der Universität Karlsruhe*, Heft 26, 1981.

10. P.K. Robertson, R.G. Campanella und A. Wightman: SPT-CPT Correlations. *J. Geotechn. Engineering, Div. Am. Soc. Civ. Engin.*, **109**: 1449–1459, 1983.

11. J.H. Schmertmann und A. Palacios: Energy Dynamics of SPT. *J. Geotechn. Engineering, Div. Am. Soc. Civ. Engin.*, **105**: 909–926, 1979.

# Two comparative field studies of the bearing capacity of vibratory and impact driven sheet piles

S. Borel, M. Bustamante & L. Gianeselli
*Laboratoire Central des Ponts et Chaussées – LCPC, Paris, France*

ABSTRACT: With the introduction of high frequency and variable eccentricity drivers, vibratory driving has become more popular because of its technical and economical performance and of considerable reduction of nuisance level. However, because of a lack of recognized method to predict the bearing capacity, vibratory driving is seldom accepted when sheet piles have to resist vertical loads. This paper presents the results of two comparative studies carried out on vibratory and impact driven sheet piles installed respectively in the very dense gravels of the Rhine river and in the marly soils with cobbles typical of the region of Paris. The instrumented sheet piles have been subjected to static loading test. By comparison with impact driven piles, the measured bearing capacity and the skin friction was 15 to 35% lower for the vibratory driven piles, whereas the penetration rate was up to 10 times faster.

## 1 INTRODUCTION

A recent survey devoted to the international practice of vibratory driving has shown that all partners involved in a foundation project (manufacturers, designers, contractors, technical supervisors and inspectors...) are worried about the real bearing capacity of vibratory driven piles (Borel & Guillaume, 2002).

This is partly due to the basic principle of vibratory driving, which consists in reducing the soil friction along the pile shaft by disturbing or liquefying the soil next to it. The estimation of the set-up of the shaft friction after it has been considerably reduced during pile installation is one of the main issues related to the use of vibratory driving for installing piles.

### 1.1 *Field data*

Published full-scale field results are very rare. The only available data concern tests carried out for specific research (Franke and Mazurkiewicz, 1975; Briaud et al., 1990) or for choosing the appropriate pile installation technique (Mazurkiewicz, 1975). Tests were sometimes run where vibratory driven piles have shown bearing capacities significantly lower than the foreseen ones (Mosher, 1990).

Available cases concern piles driven mostly in medium dense sands. For these soils, the bearing capacity of vibratory driven piles can be as much as 40% lower than the bearing capacity obtained for impact driven piles. Such reduction in capacity has been mainly attributed to a severe drop of the tip resistance.

Unlike impact driving, vibratory driving does not lead to the compaction of the soil under the pile tip, whereas the shaft friction is expected to be more or less the same for both driving techniques, and even a little higher for vibratory driven piles.

No published results could be found concerning vibratory driven piles in clays.

### 1.2 *Design practice*

Because there is no recognized method to predict the bearing capacity of vibratory driven piles, the design practice usually considers that vibratory driving is like impact driving. Generally a very conservative shaft friction is adopted and no point resistance at all is taken into account for "floating" piles.

### 1.3 *Acceptance criteria for vibratory driven piles*

The acceptance criteria are usually fixed by the project supervisor. The survey of international practice of vibratory driving (Borel & Guillaume, 2002) has shown that, for temporary works, piles are sometimes accepted without testing or monitoring. For permanent structures, no example of vibratory driven piles accepted without field load test could be found.

In most cases, a final impact driving is required. This final hammering is costly, since it decreases the installation rate and requires additional equipment on site. Nevertheless, it provides information on the bearing capacity of the pile thanks to the experience gained in impact driving (driving formula, dynamic instrumentation, etc.).

Static load tests are sometimes required to determine the real bearing capacity of vibratory driven piles. In that case, it is reported that some agencies require to perform tests on a number of piles for comparing the bearing capacity of impact driving and vibratory driving for the same depth of penetration. If necessary, the resulting additional pile length is paid by the contractor. The contractor will decide if the increase in the production rate obtained with vibratory driving justifies the extra costs induced by the tests and the necessary shaft lengthening.

All these facts show that field data are still needed when a more effective and reliable design of vibratory driven piles is wanted. This paper presents two case histories where the behavior of vibratory driven and impact driven sheet piles have been compared. Static load tests on instrumented sheet piles have then been carried out, which also made it possible to obtain the load transfer down the sheet pile.

## 2 SHEET PILES DRIVEN INTO MARLY SOILS WITH COBBLES

The first case history is relative to the construction of a retaining wall for a 6-storey building in Cachan, near Paris. The wall has been designed for a 6 meter deep excavation for a two level basement. It was made out of L2S sheet piles, 9 meters long. The excavation perimeter is approximately 140 meters long.

### 2.1 Scope of the tests

When working the project up, the contractor proposed to found the external structure of the building on sheet piles. The working load is 360 kN per meter of wall length.

This solution was considered as an interesting alternative compared with a strip footing all along the perimeter of the excavation. It was considered to be attractive by the architect and the owner because of cost savings. The technical inspector however required a full-scale load test, claiming that the calculated bearing capacity of vibratory driven sheet piles was debatable.

The Laboratoire Central des Ponts et Chaussées LCPC (French Highway Central Laboratory) was put in charge of the test and recommended to compare the behavior with the one of impact driven sheet pile. Two additional pairs of instrumented sheet piles were driven and statically loaded to failure: the first one installed with a vibratory driver and the second one using a diesel hammer (figure 1).

Figure 1. Driving the sheet piles with PTC 23HF3 and Delmag D12

### 2.2 Site investigation

The soil investigation consisted in 100 mm diameter continuous sampling and three pressuremeter PMT profiles. As shown in figure 2, the soils are clayey sand and gravely sand down to 5 meters, overlying marly soils with cobbles down to 14 meters. These last soils named "*marnes et caillasses*" are typical for the Paris area. They are a stratified formation alternating:
– dolomite cobbles layers up to 10 cm thick;
– plastic marly layers up to 50 cm thick.
The water table was found, in the present study, one meter below the ground level.

### 2.3 Observations during driving

The vibratory driven sheet piles were installed using a PTC 23HF3 high frequency driver, powered by a CAT350 (table 1).

The PTC driver proved to be extremely efficient through the *marnes et caillasses*: the sheet piles were driven from 5 m to 9 m with a mean penetration rate of 1.2 m/min. Note that the retaining wall was located at about 15 meters from an existing residential building. No complaint about vibration or noise has been recorded during the driving works.

Table 1. PTC 23HF3 main characteristics

| | |
|---|---|
| maximum frequency | 2300 rpm / 38 Hz |
| eccentric moment | 23 kgm |
| maximum centrifugal force | 1360 kN |
| maximum amplitude | 20 mm |
| vibrating weight | 2300 kg |
| total weight | 3900 kg |
| powerpack | 246 kW / 335 HP |

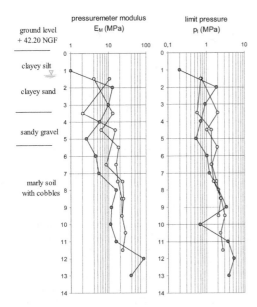

Figure 2. Soil profile at Cachan site

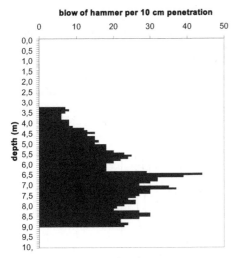

Figure 3. Driving record, Cachan site

Table 2. Observations on capacity of sheet piles

|  | Impact | Vibro | Ratio V/I |
|---|---|---|---|
| Limit load $Q_u$ | 1100 kN | 700 kN | 0.65 |
| Creep load $Q_c$ | 800 kN | 500 kN | 0.65 |
| Service load $Q_N$ | 570 kN | 360 kN | 0.65 |
| Settlement after 30 min. | 4.1 mm | 6.2 mm | 1.50 |

Figure 3 shows the driving record obtained for the impact driven pair of sheet piles. Note that the mean penetration rate was 10 cm/min, which is much slower than the rate of vibratory driving. Vibrations and noise were observed during impact driving.

### 2.4 *Static load tests*

The two sheet piles were loaded in compression to failure, after a 20 days delay, according to a Maintained Load procedure. Each load step was applied during 30 minutes using a LCPC jack. The load was checked with a 1500 kN load cell and the displacement was measured with the help of 2 linear potentiometer linked to fix reference beams. The loading frame consisted of a reaction beam supported by 10 JSP3 sheet piles driven 12 m deep (figure 4).

Figure 4. View of the piles and the loading frame, Cachan site

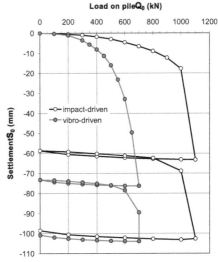

Figure 5. Load-settlement curves, Cachan site

The load settlement curves are shown on figure 5. Plunging failure was obtained for both pairs of sheet

169

piles. The vibratory driven sheet piles show a lower bearing capacity and a rather softer general response (table 2).

The measurement of the load distribution along the sheet piles was obtained through removable LPC extensometers (Bustamante & Doix, 1991). The extensometers defined 8 sections, each 1 m long. The load distributions are shown on figures 6 and 7.

The difference in bearing capacity is mainly due to the point resistance, which in the case of the vibratory driven sheet piles is only 25 % of the impact driven ones (table 3). Such a severe drop in the point resistance can be explained by differing installation techniques. It is possible also that, due to the alternation of marly and calcareous layers, the soil is stiffer at the point level for the impact driven sheet piles.

The measured skin resistance $Q_{s,u}$ is in good agreement with the values calculated using the French design code (Fascicule 62 – Titre V) yielding $Q_{s,u} = 770$ kN (for clayey soils) or $Q_{s,u} = 1075$ kN (for marly soils).

The tests have shown that the vibratory driven sheet piles were able to support the load of the building with a 10 year settlement of 10 mm, a value considered to be satisfactory.

Table 3. Limit load, skin friction & point resistance

|  | Impact | Vibro | Ratio V/I |
|---|---|---|---|
| Limit load $Q_u$ | 1100 kN | 700 kN | 0.65 |
| Skin friction $Q_{s,u}$ | 775 kN | 625 kN | 0.80 |
| Point resistance $Q_{p,u}$ | 325 kN | 75 kN | 0.25 |

# 3 SHEET PILES DRIVEN INTO DENSE GRAVELS

The second case history is relative to the construction of a retaining wall for an inlet lock built across a dike along the Rhine River in Erstein (Alsace, France). The lock is intended to regulate the river flow in case of a water rise by flooding a polder (figure 8).

The wall was made out of PU20 and L3S sheet piles 12 meter long. It was constructed to support a 7 meter deep excavation, but also to resist water uplift buoyancy when the inlet structure is empty.

## 3.1 *Site investigation*

The soil investigation consisted in a CPT test and two pressuremeter PMT profiles. As shown in figure 9, the site comprises 1.6 m of gravely fill overlying the dense gravels of the Rhine River. According to the Unified Soil Classification System USCS, these alluvia are of GP type. Their Grain size distribution is shown on figure 10. The water level follows the Rhine River level.

## 3.2 *Observations during driving*

Two pairs of PU20 sheet piles were installed at a distance of 6 meter using respectively:
- an ICE 815 standard frequency vibratory driver (table 4) ;
- a Delmag D30 diesel hammer.

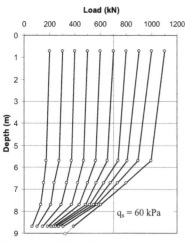

Figure 6. Load distribution versus depth for impact driven pile, Cachan site

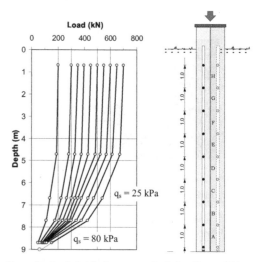

Figure 7. Load distribution versus depth for vibratory driven pile, Cachan site

Figure 8. Sheet pile at Erstein inlet lock

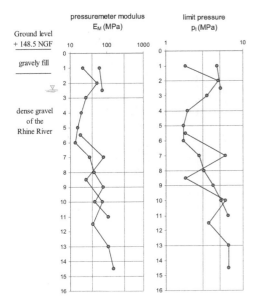

Figure 9. Soil profile at Erstein site

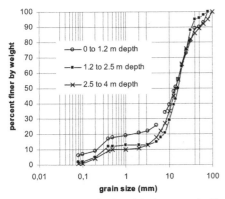

Figure 10. Grain size distribution of the dense gravel at Erstein

Table 4. ICE 815 main characteristics

| maximum frequency | 1570 rpm / 26 Hz |
|---|---|
| eccentric moment | 46 kgm |
| maximum centrifugal force | 1250 kN |
| maximum amplitude | 18 mm |
| vibrating weight | 5050 kg |
| total weight | 8550 kg |
| powerpack | 384 kW / 522 HP |

The vibro-driver ICE proved to be extremely efficient for the penetration into gravels. Each sheet pile was driven to a depth of 11 m with a mean penetration rate of 10 m/min under the total weight of the pile and the vibrodriver (i.e. no crane uplift being applied).

Figure 11 presents the driving record obtained for impact driven sheet piles. The mean penetration rate of 1 m/min is again 10 times slower than the one of the vibratory driven sheet piles.

Because the inlet lock has to be constructed across a dike, the project supervisor required to control the vibration velocities during driving. At a distance of 3 m from sheet piles, a ground velocity up to 40 mm/s was measured for a number of blows during impact driving, whereas the geophone trigger level of 10 mm/s was not reached during vibratory driving.

### 3.3 *Static load tests*

Each pair of sheet piles was statically pulled out after a delay of one week, according to a Maintained Load procedure, each load step being applied during 30 minutes. The displacement was measured with the help of 4 linear potentiometers. As shown on figure 12, the loading frame consisted of a reaction beam propped by a stabilizer system supported on stacks of sleepers.

Removable LPC extensometers defining 8 measurement sections were used to measure the load transfer down the sheet pile (figure 13).

The load settlement curves are shown on figure 14. Two loading cycles were applied. A plunging failure was observed for the vibratory driven sheet piles for a limit load $Q_u = 1650$ kN, whereas the impact driven pair was loaded up to 2250 kN. The vibratory driven pair of sheet piles shows a lower bearing capacity (table 5), but the general response is very similar up to working load.

The instrumentation made it possible to determine the load distribution during the test (figures 15 and 16) and the mobilization curves of the unit skin frictions (figures 17 and 18).

No significant residual stresses have been measured for the vibratory driven piles whereas, for the impact driven one, an apparent tension load of about 750 kN was measured under the point after the first load cycle.

For the vibratory driven sheet piles, the mean skin resistance is more than 25% lower than for the impact driven ones. The full mobilization of frictional resistance is observed for a displacement greater than 20 – 40 mm. This value is significantly larger than the usual values ranging between 2 and 10 mm for common soils.

The measured skin resistance is in good agreement with the values given by the French design code (Fascicule 62 – Titre V), i.e. $q_s = 40$ kPa for gravels 4 to 6 m deep and $q_s = 60$ kPa below 7 m.

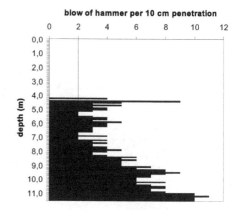

Figure 11. Driving record, Erstein site

Figure 13. Inserting of the removable LCPC extensometers

Figure 12. View of the loading frame (Erstein)

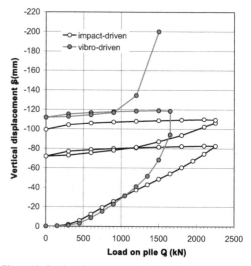

Figure 14. Load-settlement curves, Erstein site

172

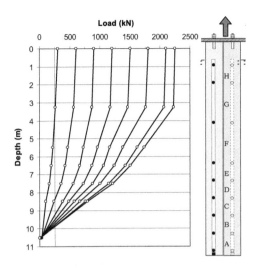

Figure 15. Load distribution versus depth for impact driven pile (Erstein – 2nd cycle)

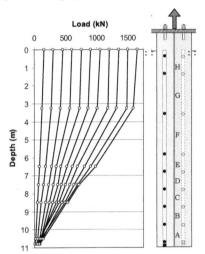

Figure 16. Load distribution versus depth for vibratory driven pile (Erstein - 1st cycle)

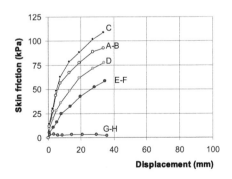

Figure 17. Curves of unit skin friction versus displacement at different depths (impact driven pile at Erstein – 2nd cycle)

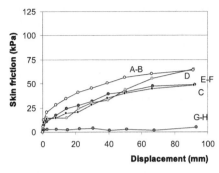

Figure 17. Curves of unit skin friction versus displacement at different depths (impact driven pile at Erstein – 2nd cycle)

Figure 18. Curves of unit skin friction versus displacement at different depths (vibratory driven pile at Erstein- 1st cycle)

## 4 CONCLUSIONS

Comparative field studies of the bearing capacity of vibratory and impact driven sheet piles have been performed on two sites. The main conclusions for vibratory driven piles can be drawn, as follows:
– in dense gravels, the measured skin friction is more than 25% lower than for impact driven sheet piles;
– in marly soils with cobbles (*marnes et caillasses*), the observed point capacity and skin friction were respectively 75% and 15% lower.

Even with these reductions in bearing capacity, vibratory driving was considered to be satisfactory for both projects, where sheet piles had to carry vertical loads. Using vibratory driving has allowed higher penetration rate, with less noise and vibration transmitted to the surroundings.

The authors are of the opinion that for common design practice and for "floating" piles, the point resistance should not be taken into account when designing the bearing capacity of vibratory driven sheet piles. Two main reasons support this view:
– available measured point resistances are usually relatively small compared to the total bearing capacity;
– upward and downward vertical displacement induced to the pile during vibratory driving does not guarantee a good contact of the pile tip with the bearing layer. As observed in practice, sheet piles are often lifted up after their installation, in order bring into alignment at the top.

## ACKNOWLEDGEMENT

The authors would like to thank piling contractors LEDUC and DURMEYER who undertook the task of installing the sheet piles. These tests have been supported by the owner Voies Navigables de France VNF Strasbourg. LRPC Strasbourg, LROP Trappes and ISPC Arbed are also acknowledged for their support on site.

Roger Frank is acknowledged for his valuable remarks during reviewing the paper.

## REFERENCES

Borel, S. & Guillaume, D. 2002. Present issues of vibratory driving in urban areas. *Proceedings 2nd International Conference on Soil-Structure Interaction in Urban Civil Engineering, Zurich, 7-8 March 2002.* Rotterdam: Balkema.

Briaud J.-L., Coyle H.M., Tucker L.M. 1990. Axial response of three vibratory and three impact driven H piles in sand. *Transportation Research Record. No. 1277. pp. 136-147.*

Bustamante, M. & Doix, B. 1991. A new model of LPC removable extensometer. *Proceedings 4th international conference on piling and deep foundation, Stresa, 7-12 April 1991.* Rotterdam: Balkema.

Franke E., Mazurkiewicz B. 1987. The influence of the pile installation method on its bearing capacity. *Proceedings, 8th Polish national conference on soil mechanics and foundation engineering. vol. 2, pp. 533-538.*

Mazurkiewicz B. 1975. The influence of vibration of piles on their bearing capacity. *Proceedings, 1st Baltic Conference on Soil Mechanics and foundation engineering, Gdansk, section 3, pp. 144-153.*

Mosher R.L. 1990. Axial capacity of vibratory-driven piles versus impact driven piles. *Transportation Research Record. No. 1277. pp. 128-135.*

Règles techniques de conception et de calcul des fondations des ouvrages de génie civil. *Cahier des clauses techniques générales applicables aux marchés publics de travaux. Fascicule 62 – Titre V 1993.* Ministère de l'équipement, du logement et des transports. Paris: Textes officiels (in French).

# Pile Load Tests of Large Diameter Battered Steel Pipe Piles Constructed in an Offshore Area

J. H. Lee, D. D. Seo, S. Y. Kwon, J. H. Won and Y. I. Baek
*Daelim Industrial Co., Ltd., Seoul, Korea*

ABSTRACT: Battered piles were driven for the foundation of Gwang-An grand bridge in Pusan, Korea. The bridge foundation consists of 8 to 12 piles and each one is composed of two materials. The outer part of the pile consists of a 28 mm thick with a diameter of 2500mm while the inner part is filled with concrete. The construction of pile is performed through three steps. First, the steel pipe is driven using the hydraulic vibrating hammer on the jack-up barge. The next step is digging out the soil and rock about 6 meters deeper than the toe of the steel pipe using RCD equipment. The final step is pouring concrete through the tremie pipe. Each pile is designed to bear 1500 ton and the pile end is to be located on the soft rock. The end of 3 piles are, however, located on the weathered rock because of the workability of the RCD equipment. In order to verify the allowable bearing capacity of these piles, several pile load tests were carried out.

## 1 INTRODUCTION

A number of 11.3° battered piles of 2500 mm diameter were driven in order to support Gwang-An grand bridge in Pusan, Korea. Those piles were composed of two materials, steel pipe pile and concrete. Since this bridge is located on the sea bed, jack-up barges with 36 m long legs were required to drive the piles and other construction equipments such as hydraulic vibrating hammer and RCD equipment are used.

Each pile was designed to bear 1500 tons and the pile toe was to be located on the soft rock whose strength was above 700 kgf/cm². Some pile toes were, however, located on the weathered rock whose strength was 180 kgf/cm² and no more excavation could be done because of the workability of RCD equipment. In order to verify the allowable bearing capacity of these piles, several pile load tests were carried out. End bearing capacity of the pile was measured by the plate bearing test within the steel pipe pile using the skin friction of steel pipe pile as a kentledge load and dynamic load tests were also applied. Since the piles were inclined and constructed on the offshore area, special test instruments were developed.

## 2 CONSTRUCTION OF PILE FOUNDATION

### 2.1 Geological Conditions

The construction site is on the sea bed and the geological condition falls into 2 parts. Upper part, near the ground surface, consists of various soil layers such as silty sand, sand, gravel and clay of which the thickness is about 6 to 8m. Lower weathered part, whose thickness is about 18 to 30m, consists of residual soil, weathered rock, and the base rock.

### 2.2 Construction sequence

Using the hydraulic vibrating hammer described in Table 1 and Figure 1, the steel pipe pile was driven first. After the pile driving, inner soil of the pile was removed by the RCD equipment. The 2200mm diameter concrete pile was constructed without steel pipe from the bottom end of the steel pipe pile to the base rock. Figure 2 shows the schematic diagram of the pile construction sequence.

Table 1. Specification of the vibrating hammer.

| | |
|---|---|
| Excentric moment | 105 kg-m |
| Max. frequency | 1350 rpm |
| Max. centrifugal force | 215 ton |
| Max. line pull | 120 ton |
| Max. amplitude | 25 mm |
| Vibrating weight | 8400 kg |
| Total weight without clamp | 13800 kg |

Figure 1. Hydraulic vibrating Hammer

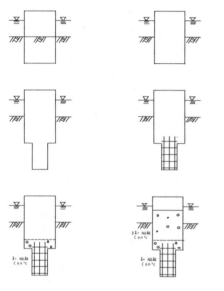

☐  Pile driving to the weathered rock using the vibrating hammer
☐  Digging out the soil using RCD
☐  Digging out  the soil to the base rock without steel pipe
☐  Rebar cage installation
☐  1st concrete pouring
☐  2nd concrete pouring

Figure 2. Schematic diagram of pile construction sequence

# 3  PILE LOAD TESTS

## 3.1  Design of Pile Foundation

In order to resist the various loads of the bridge, large battered pile were designed. The composite pile was made of steel pipe filled with concrete. The inner diameter of steel pipe was 2500 mm and the thickness was 28 mm. The pile toe was designed to be located on the base rock whose strength was more than 700 kgf/cm$^2$ and the vertical service load of one pile was 1500 tons. Observing the rock piece of RCD excavation indicated that the toe of piles were located on the weathered rock as shown in figure 3.

No more excavation, however, was possible because of the workability of the RCD equipment. Both static and dynamic pile load tests were planned to verify the stability of the piles.

## 3.2  Static Load Test-Plate Bearing Test

Since the piles were inclined and constructed on the offshore area, special test instruments were developed for the safety of the load tests. These are shown in Figures 4 and 5. The inner pile of 812 mm diameter was installed in the steel pipe pile to verify the end bearing capacity. The end of inner pile was closed by the circular plate of 820 mm diameter and 60 mm thick. The skin friction of outer pile was utilized as kentledge weight and the test load was applied by the hydraulic jack.

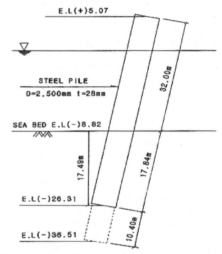

Figure 3. Piles on the weathered rock

Figure 4. Inner pile installation

176

Figure 5. Loading frame

Figure 7. Attachement of gauges

Among the test piles located on the weathered rock, two piles were already filled with concrete and only one pile was empty. Therefore, the plate bearing test was applied to verify the end bearing capacity of this empty pile. Figure 6 shows the end bearing capacity is 143.4 ton.

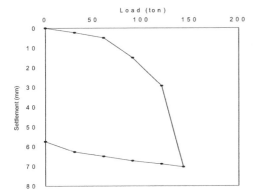

Figure 6. Test result of the plate bearing test

The test result should be modified because of the overburden pressure in the plate bearing test. The theory proposed by Hansen (1970) was adopted in order to consider the confining effect and the depth factor. From this consideration, the end bearing capacity was calculated as 1361.9 ton. The friction angle of 40° and the ratio of depth/diameter of 13 were assumed in this calculation.

### 3.3 Dynamic Load Test

The dynamic load tests were carried out to evaluate the skin fiction. The dynamic load tester was made by PDI. As shown in Figure 7, the gauges (4 accelelometers & 4 strain gauges) were attached on the steel pipe pile.

For the driving load, an a S-280 IHC hydraulic piling hammer was used. The ram weight of this hammer was 13.5 ton and the maximum driving energy was 280 kJ. In order to install this hammer on the pile, the cap of hammer was newly made as shown in Figure 8.

Figure 8. Pile driving hammer

The test pile was divided by 3 parts, only steel, steel and concrete and only concrete pile. All of these parts had different characteristics and so the obtained test result had to be corrected. For the accurate test, the gauges could be attached on the concrete because the concrete was so massive that it governed the loading behavior of pile. Since the piling condition, however, did not permit attaching the gauges on the concrete, the impedance of pile was remodeled during the analysis of test result. In the analysis of wave matching, concrete section was converted to the steel one considering the elasticity and section modulus.

The results of the dynamic test showed the skin friction was 4530 ton. Although it was enough to resist the bridge load, this result was obtained by the assumption of the concrete pile impedance. Moreover, the dynamic load test result showed the skin friction was distributed on the steel surface above the pile composed of concrete only. Therefore, it is necessary to investigate the skin friction of the concrete part.

## 3.4 *Application of Point Load Tests*

In order to investigate the skin friction of the concrete part of test pile, point load test was applied to the rock pieces from the RCD excavation. The test result showed the unconfined strength of the rock was min. 180 kgf/cm$^2$. Haberfield & Seidel (1996) proposed the evaluation method of the skin friction of rock socketed pile and the relationship between the unconfined strength and the skin adhesion as shown in Figure 9. Considering this figure and unconfined strength, the unit skin friction was 0.3~0.5 MPa. From this result, the skin friction of the concrete part was at least 3000 tons.

rock. That is to say, the working load of the foundation is somewhat overestimated.

## 5 REFERENCES

– C.M. Haberfield & J.P. Seidel (1996), "A new design method for drilled shafts in Rock," Proceedings of the Sixth International Conference on Piling and Deep Foundations, Bombay.
– J.B. Hansen (1970), "A revised and extended formula for bearing capacity," Danish Geotechnical Institute Bul, No.28, 21pp.

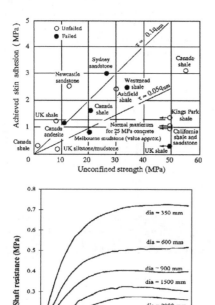

(a) Effect of unconfined strength
(b) Effect of pile diameter

Figure 9. Parameters affecting the skin friction

## 4 CONCLUSION

Although there were differences between the design and the construction, all of the test results showed the piles could resist the working load of the bridge without the additional excavation down to the soft

*Vibratory prediction event*

*Event and predictions comparison*

*Vibratory Pile Driving and Deep Soil Compaction - TRANSVIB2002,*
*Holeyman, VandenBerghe & Charue (eds.), © 2002 Swets & Zeitlinger, Lisse, ISBN 90 5809 521 5*

# Full-scale behavior of vibratory driven piles in Montoir

S. Borel & L. Gianeselli
*Laboratoire Central des Ponts et Chaussées*

D. Durot & P. Vaillant
*Rincent BTP*

L. Barbot
*EEG Simecsol*

B. Marsset & P. Lijour
*Port Autonome de Nantes Saint-Nazaire*

ABSTRACT: Within the framework of the French National Project on vibratory driving, it was decided to investigate the full-scale behavior of vibratory driven piles in the harbor of Montoir, located at the estuary of the Loire River. The subsoil consists of alluvial sandy deposits overlying sandy mud clay. Two 339 mm closed-end tubes, 31 m long, were vibratory driven to 19 m depth where the refusal was observed. To investigate dynamic resistance, one of the tubes was impact driven for the last meter by a diesel hammer. The tubes were both instrumented with accelerometers and strain gauges positioned at the top and at the pile point. The entirely vibrodriven pile was statically loaded after a delay of 3 weeks. In this paper, the authors report on the site investigation, the observations made during the driving and the results of the static loading test. Bearing capacity of the vibratory driven pile is compared to previous tests on CFA and impact driven piles.

## 1 INTRODUCTION – SCOPE OF THE TESTS

The new container terminal TMDC 4 in Montoir-de-Bretagne (France) is located on the estuary of the Loire River, close to the Saint-Nazaire Bridge. The quay is 250 m long and 45 m wide. It is made of a precast concrete structure, supported by end bearing piles driven to refusal against a layer of gneiss investigated at a depth of approximately 48 m (figure 1).

At the back of the quay, the fill is up to 8 m high and 25 m large. To reduce the excessive consolidation settlements, which were anticipated, it was proposed to support the embankment on a network of floating piles. As more than 500 floating piles were to be constructed, the harbor authority decided to investigate the real full-scale behavior of different piling techniques.

In 1999, two piles were installed on site and statically loaded to failure:
- A CFA pile Ø600 mm 18.3 m long;
- A closed-end steel tube Ø339 mm, driven to 31 m depth by a diesel hammer Delmag D22.

As shown on figure 2, the foundation finally chosen consisted of 230 CFA piles of the Starsol type, Ø540 mm with a length of 32 m. The piles were spaced at 3.5 m center with a regular pattern. They were installed at the beginning of 2001.

Figure 1. The terminal during construction

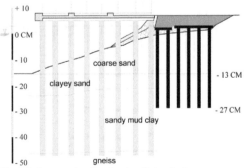

Figure 2. Cross section showing the piled embankment

In connection with the French National Project on vibratory driving, the harbor authority decided to investigate the behavior of closed-end steel tubes (diameter Ø339 mm, 14 mm thick) very similar to the pile which was impact-driven in 1999. It was anticipated to vibratory drive the tube down to 32 m in only one piece using a vibratory driver ICE 815 (table 1). But the refusal was observed at 18.90 m only.

A second tube of the same type was vibratory driven down to 17.90 m and then impact driven using a Delmag D12 (table 2) until 18.90 m.

Table 1. ICE 815 vibrator main characteristics (with clamps)

| maximum frequency | 1570 rpm / 26 Hz |
|---|---|
| eccentric moment | 46 kg.m |
| centrifugal force | 1250 kN |
| maximum amplitude | 20 mm |
| vibrating weight | 5500 kg |
| total weight | 8000 kg |
| powerpack | 384 kW / 522 HP |

Table 2. Delmag D12 main characteristics

| ram weight | 1280 kg |
|---|---|
| blow rate | 36 to 52 blow/min |
| max. potential energy per blow | 2090 to 4330 kg.m |

The layout of the test site is shown on figure 3 where the first piles tested in 1999 are called "CFA" and "impact". The entirely vibratory driven pile is designated by "vibro" and the pile which was impact driven for the last meter is designated by "vibro + impact".

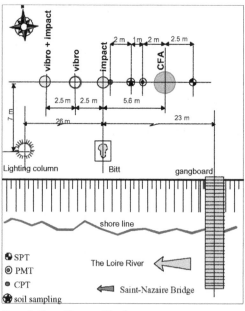

Figure 3. General layout of the site

## 2 SITE INVESTIGATION

The ground level is located at 7.5 m above the lower level of the sea (+ 7.5 CM). The water table follows the level of the Loire River, which is subjected to tides.

The subsoil was investigated with a 90 mm diameter continuous sampling boring, PMT pressuremeter, CPT and SPT profile (figures 4, 5 and 6). As shown in table 3, the subsoil consists of:
– from 0 to around 5 m, a fill made of coarse sand with gravels and sometimes blocks;
– from 5 to 22.5 m, alluvial deposits with lenses of sandy clay.
– from 22.5 to 36 m, mud clay with fine sand layers of some millimeters. This soil is locally named "jalle".

Typical grain distributions are shown on figure 7.

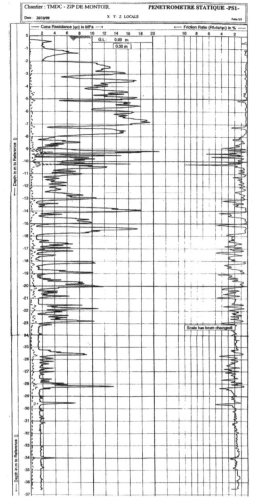

Figure 4. CPT profile

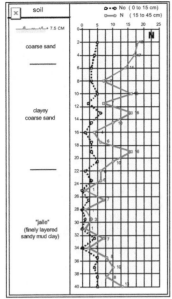

Figure 5. SPT profile

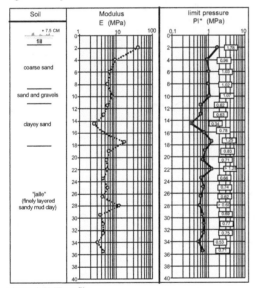

Figure 6. PMT profile

Table 3. Soil Stratigraphy in Montoir

| depth below ground surface (m) | Soil description |
|---|---|
| 0 to 4.5 | brown coarse sand with gravels |
| 4.5 to 8.5 | gray-black sand |
| 8.5 to 10 | block |
| 10 to 13 | slightly clayey coarse to fine sand |
| 13 to 22.5 | alternation of coarse to fine sand layers (40-100 cm thick) and gray-black sandy clay layers (10-30 cm thick) |
| 22.5 to 36 | finely interbedded (1 to 20 cm) alternation of fine sand and mud clay (locally named "jalle") |

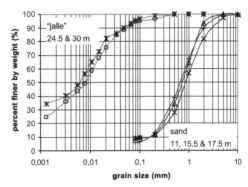

Figure 7. Grain size of the sandy alluvial deposit and the "jalle"

## 3 INSTRUMENTATION OF THE PILES

The vibratory driving was to be monitored using continuous record of :
- the penetration rate;
- the acceleration and the stresses at the top and at the pile point;
- the uplift load applied by the crane;
- the vibrations transmitted to the ground.

The piles were marked every 10 cm, which made it possible to deduce the penetration rate from video recording (figure 8).

Both tubes were instrumented with accelerometers and strain gauges positioned at the top and at the pile point on two diametrically opposite measuring sections.

Strain gauges were TML-FCA-6-11 rosette type with temperature compensation. Accelerometers were Entran EGCS-D2-50 type with a scale range of 50 g, a sensitivity of 0.26 mV/g/V and a passing band of 600 Hz at 0.5 dB (figure 9).

The uplift load applied by the crane was measured with two strain gauges similar to the ones used for the tubes and placed directly on the crane hook (figure 10).

All the sensors were connected to a central HBM MGC-Plus type data acquisition system at 1200 Hz.

As shown on figure 11, the instrumentation and the cables were protected by corner irons welded on the pile. Two additional steel tubes were soldered to receive the LPC removable extensometers during the static loading test (Bustamante & Doix, 1991). At the bottom, a flat plate of steel was soldered in order to protect the instrumentation.

As reported in table 4, the total steel surface exceeded the net tube section.

Table 4. Steel surface (cm²)

| | tube | iron corners | extensometers | total |
|---|---|---|---|---|
| pile toe | 903 | 171 | 56 | 1130 |
| pile cross-section | 143 | 4 | 13 | 160 |

Figure 8. Painted marks on piles during driving

Figure 9. Strain gauges and accelerometers at bottom of pile 1

Figure 10. Gauges placed on hook

Figure 11. Protection of the instrumentation at the pile toe using corner iron

Figure 12. Velocity transducer

The vibrations transmitted to the ground were measured by 3 geophones (Marks Products L-4C-3D) installed at 6 m, 12 m and 18 m from the pile. The velocity transducers measured the velocity in three orthogonal directions at an acquisition frequency equal to 1000 Hz (figure 12).

## 4 OBSERVATIONS DURING INSTALLATION

Both piles were driven the August 30, 2001 in a single section of 32 m (figure 13). The first pile was entirely vibratory driven. The ICE driver proved to be very efficient in the clayey coarse sand, from 6 to 14 m depth, where the maximum penetration rate reached 6 m/min. Following an interruption of 3 minutes the penetration speed decreased when the pile entered into a ground alternating clayey and sandy strata below 14 m deep. The vibratory driving was finally stopped at 18.9 m deep after the pile had penetrated only 20 cm in 2 minutes, which was interpreted as refusal.

As shown on figures 14 and 15, the driving record of the second pile is very similar to that of the first one down to 17.9 m. The second pile was then impact driven using a Delmag D12 at a penetration rate of 52 to 54 blow/min and a penetration speed of 15 to 20 cm/min.

The uplift load from the crane estimated using the strain gauges placed on the hook is shown on figure 16. Strain gauges placed on the top of the pile lead to very similar uplift loads. Three different ways of driving were observed:
- from 0 to 5 m, the pile was held back while it penetrated the coarse sandy fill at a mean speed of 0.4 m/min (max speed 1.2 m/min) ;
- from 5 to 13 m, the pile was driven through slightly clayey coarse to fine sand at a mean speed of 2 m/min (max speed 6 m/min). The pile was first slightly held back;
- from 15.5 to 17.9 m, the full weight of the pile plus the vibratory driver was applied to penetrate the alternation of sand and clay. But the mean speed decreased down to 0.4 m/min (max speed 1.2 m/min).

During vibratory driving, data shows periodic signal at 23 Hz at the beginning of driving decreasing to 21.5 Hz at the end of the penetration.

The data recorded during the vibratory driving of the first pile are limited. Accelerometers placed at head did not work during the whole test. Others data acquisition sensors give incomplete information. Therefore, in this paper are reported only the data recorded during the vibratory driving of the second pile, which was impact driven for one supplementary meter.

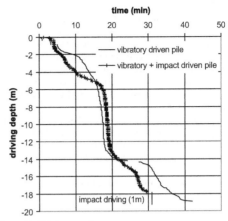

Figure 14. Driving record

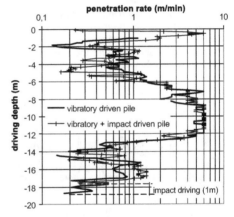

Figure 15. Penetration rate

Figure 13. Driving the piles

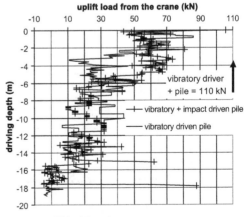

Figure 16. Uplift load from the crane

185

## 4.1 Accelerometers measurements

Figure 17 shows a typical acceleration signal measured at a penetration depth of 11 m, when the mean penetration speed reached 6 m/min. Spectral analysis shows peak acceleration at 22.5 Hz corresponding to the driver frequency (figure 18). Harmonic peak acceleration of pile toe is observed at 45 Hz corresponding to the flattening of the pile toe displacement at the end of the downward movement (see figure 19).

The vertical displacements were calculated by a double integration of the accelerations. As shown on figure 19, the head and toe displacements are in-phase with an amplitude of 11 mm at the head and 13 mm at the toe level.

This amplitude is similar to the theoretical amplitude of 10 mm calculated for a free profile using :

$$2 \, eM_e / (M_p + M_v) \tag{1}$$

where $eM_e$ is the eccentric moment, $M_p$ is the pile weight and $M_v$ is the vibrating weight.

Peak accelerations versus depth are shown on figure 20. Maximum and minimum head accelerations are similar, whereas minimum toe acceleration is lower, corresponding again to the flattening of the toe displacement at the end of the downward movement.

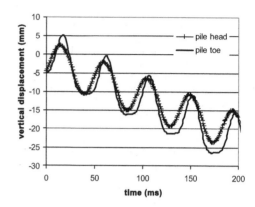

Figure 19. Displacement at a penetration of 11 m depth

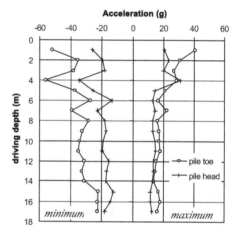

Figure 20. Peak accelerations

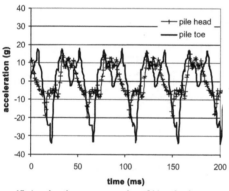

Figure 17. Acceleration at a penetration of 11 m depth

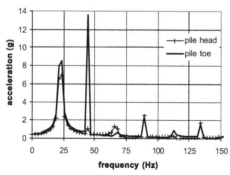

Figure 18. Spectral analysis of the acceleration at 11 m depth

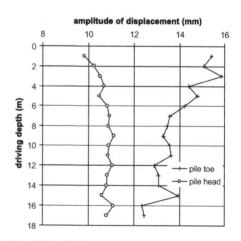

Figure 21. Amplitude of displacement

As shown on figure 21, the amplitude of the vertical displacement remains between 10 mm and 15 mm for all the penetration phases. The amplitude of the toe displacement is apparently higher than head displacement.

### 4.2 Strain gauges measurements

At the same penetration depth of 11 m, the strain measured at the head of the pile is shown on figure 22. The two gauges measured different signal shapes and amplitudes. For some other penetration depth, the signals are even in opposite phase. Such phenomena can be induced by the superposition of a flexion to the traction/compression force at the top of the pile. Difference in amplitude can also be justified by a non-axially force.

The values shown on figure 23 gives an amplitude of approximately 600 kN, which is lower than the theoretical force of 1000 kN calculated for a free profile using :

$$2 F_c M_p / (M_v + M_p) \qquad (2)$$

where $F_c$ is the centrifuge force, $M_p$ is the pile weight and $M_v$ is the vibrating weight. As shown on figure 24, the theoretical amplitude of 1000 kN was observed at the very beginning of the vibratory driving.

It seems that the strain gauges located at the toe of the pile were damaged after the pile passed through the blocks at 9 m depth.

### 4.3 Vibration transmitted to the ground

Peak particle velocities were measured at 6 m, 12 m and 18 m from the pile. The vertical, radial, transverse and resultant peak velocities are plotted versus penetration depth on figures 25, 26 and 27.

Measured vibrations are of a complex nature. Transverse movements were most important at 6 m and exceeded 15 mm/s whereas vertical velocities didn't reached 4 mm/s at that distance from the pile. On the contrary, vertical velocities were predominant at 12 m and 18 m from the pile.

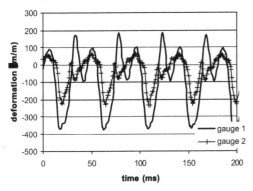

Figure 22. Head strain at a penetration of 11 m depth

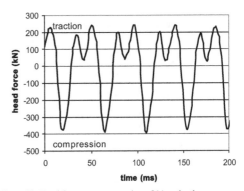

Figure 23. Head force at a penetration of 11 m depth

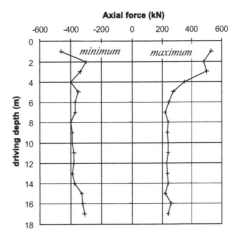

Figure 24. Peak axial head force

The peak velocities were higher when the pile was penetrating down to 7 m depth through the upper sandy strata. Then vibrations decreased by a factor 2 when the penetration speed increased through the alternation of sandy and clayey layers down to 12 m. At the end of driving, after the driving speed was decreased, the level of vibrations increased again.

Spectral analyses are shown on figures 28, 29 and 30. Harmonic peak velocities are sometimes higher than the peak velocity at fundamental frequency corresponding to the vibratory driver excitation. This is particularly clear for horizontal velocities. Because the pile is slender, it is possible that it was subjected to lateral vibration. Resonant frequencies of the pile have been investigated. Eigen frequencies of approximately 50 Hz (1st mode) and 150 Hz (2nd mode) were estimated for vertical movement of a free profile; whereas under lateral movement eigen frequencies were much lower: 1 Hz (1st mode), 6 Hz (2nd mode) and 17 Hz (3rd mode). It is therefore possible that driver's frequency repeatedly matched resonant lateral pile frequency, causing important horizontal vibrations transmitted to the soil.

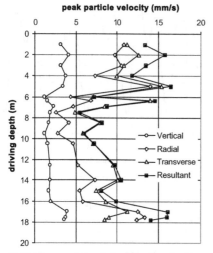

Figure 25. Maximum particle velocity at 6 m from the pile

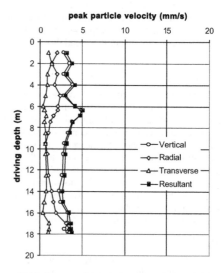

Figure 27. Maximum particle velocity at 18 m from the pile

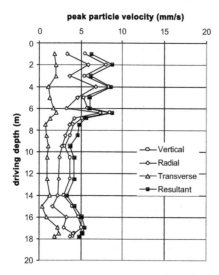

Figure 26. Maximum particle velocity at 12 m from the pile

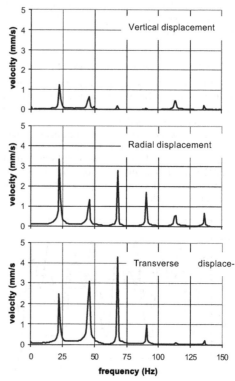

Figure 28. Spectral analysis of the velocities measured at 6 m from the pile (penetration depth = 4.5 m)

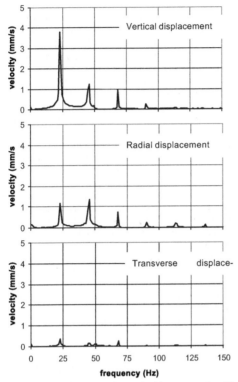

Figure 29. Spectral analysis of the velocities measured at 12 m from the pile (penetration depth = 4.5 m)

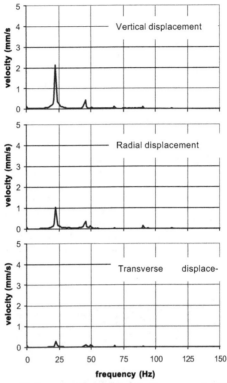

Figure 30. Spectral analysis of the velocities measured at 18 m from the pile (penetration depth = 4.5 m)

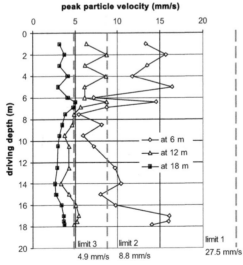

Figure 31. Maximum particle velocity compared to DIN 4150 limit criteria

Peak particle velocities can be compared to the recommendations of allowable vibration on buildings given in codes and standards. Usually, the tolerable limit particle velocity increases with higher frequency. Lower limits are recommended for continuous vibration. For a frequency equal to 25 Hz, the DIN 4150 standard gives following limits of foundation vibrations:

- $v_1$ = 27.5 mm/s for structures used for business, industrial buildings and similarly designed structures;
- $v_2$ = 8.8 mm/s for living quarters, family houses and building used for housing purposes;
- $v_3$ = 4.9 mm/s for structures which are particularly sensitive to vibrations.

As shown on figure 31, the DIN 4150 criteria are satisfied for living quarters at 12 m from the pile. For sensitive structures, the criteria are not exceeded at 18 m from the pile.

## 5 PILE STATIC LOADING TEST

The vibratory driven pile was loaded in compression to failure on September 19, 2001 after a 20 day delay, following a Maintained Load Test procedure. Each 50 kN load step was applied during 30 minutes. The load was controlled using a 1500 kN load cell, and the vertical displacement was measured with 2 linear potentiometers attached to reference beams (figure 32).

The loading frame consisted of a reaction beam (2 HEB 800) supported by two tubes Ø339 mm: The first was impact driven in 1999 to 31 m deep; The second is the 2001 test tube which was vibratory driven to 17.80 m and impact driven for one supplementary meter. The device was designed to load the pile up to 2000 kN.

Load settlement curve is shown on figure 33. A plunging failure was observed under a limit load $Q_u = 1250$ kN. The pile settlement exceeded 10% of the diameter with a value S = 67 mm observed after 15 minutes under the maximum load, afterward the pile was unloaded.

Figure 32. Reaction beam and loading device

Figure 36. Assembling the removable extensometers

The analysis of the time settlement curve determinates a creep load $Q_c = 800$ kN (figures 34 and 35).

In order to obtain the load distribution along the shaft, the pile has been instrumented with removable LPC extensometers (figure 36). The extensometers delimitated 10 measuring sections (called A to J). The load distribution is shown on figure 37. The mobilization curves of the unit skin frictions are plotted on figure 38.

The maximum applied load $Q_u = 1250$ kN was supported mainly by shaft friction (80%), whereas only $Q_p = 260$ kN was resisted under the pile point, which corresponds to a pressure $q_u = 2.3$ MPa. Following shaft friction $q_s$ were observed:

−  $q_s = 45$ kPa from 0 to 8 m depth (measuring level G, H, I and J) for the coarse sand characterized by a CPT resistance $q_c = 10$ to 18 MPa, a friction ration FR < 0.5 % and a PMT limit pressure $p_l = 1$ MPa;

−  $q_s = 20$ kPa from 8 to 14 m depth (measuring level D,E and F) for the clayey coarse sand characterized by a lower CPT resistance $q_c = 3$ to 10 MPa, a friction ration FR ~ 1 % and a PMT limit pressure $p_l = 0.6$ to 1 MPa;

−  a peak shaft resistance $q_s = 35$ kPa, stabilized to a residual value $q_s = 30$ kPa, from 14 to 18.9 m deep (measuring level A, B and C) for the sand getting more and more clayey with a limit pressure $p_l$ ~ 1 MPa. This layer alternates clayey lenses ($q_c = 1.5$ MPa and FR = 2 to 4%) and sandy strata ($q_c = 6$ to 18 MPa and FR ~ 1%).

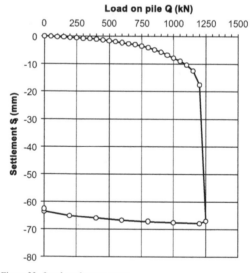

Figure 33. Load-settlement curve

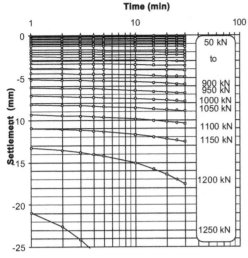

Figure 34. Time-settlement curves

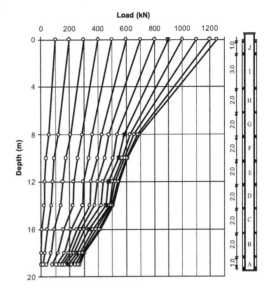

Figure 37. Load distribution versus depth

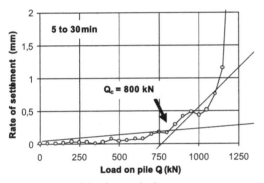

Figure 35. Determining the creep load

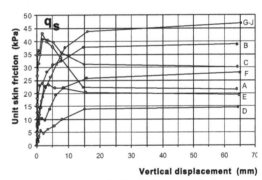

Figure 38. Curves of unit skin friction versus displacement at different shaft level

Table 5. Skin friction measured on different piles in Montoir

| depth below ground surface | CFA | vibratory driven | impact driven |
|---|---|---|---|
| 0 to 8 m | > 55 kPa | 45 kPa | 10 kPa* |
| 8 to 14 m | > 45 kPa | 20 kPa | 10 kPa* |
| 14 to 19 m | > 25 kPa | 35 kPa | 30 kPa |
| 22.5 to 36 m | | | 70 kPa |

\* very low skin friction is probably due to the slight enlargement of the pile shoe

The behavior of the vibratory driven pile can be compared to previous loading tests carried out in 1999 (Bustamante et al., 1999). In order to obtain the load distribution along the shaft, each pile has been instrumented. Measured skin friction are reported in table 5. The direct comparison is not easy because:

– the geometry of the piles differs from the Ø339 mm vibratory tube, which is 18.9 m long. The Ø600 mm CFA pile was 18.2 long and the impact driven Ø339 mm tube was 31 m long;
– the CFA pile was loaded up to 1500 kN, which was far from the failure load, as the observed settlement was 3.7 mm only under this maximum load;
– the diameter of shoe below the point of the impact driven tube was slightly larger than the diameter of the tube. This probably explains the very low skin friction measured in the sandy strata down to 14 m.

The measured end bearing resistance is intermediate between the values given by the French piling code (fascicule 62 – Titre V) for impact driven closed-end tubes driven in sands and clays with the mechanical properties investigated on site. This corresponds in fact to the alternation of sands and clays observed at the toe level.

The measured skin frictions are only 50 % to 75 % of the values estimated using the French code for impact

191

driven tubes. This relatively low bearing capacity of the vibratory driven piles compared to impact driven ones was observed on other sites (Franke and Mazurkiewicz, 1975; Briaud et al., 1990; Mosher, 1990; Borel et al., 2002). But more tests are needed for a better understanding and for a more reliable and effective design of vibratory driven piles.

## 6 CONCLUSIONS

Field behaviour of vibratory driven piles has been studied in the harbour of Montoir. Two closed end tubes were instrumented, driven and statically loaded. Even if a part of the instrumentation was damaged during the pile installation, these tests constitute a very rare and valuable case history.

As shown in table 6, a close correlation has been observed between the soil investigation, the penetration speed and the skin friction.

In connection with the tests carried out in Montoir, a blind prediction event has been organised before installing the pile. Predictions were compared to the field observations during the International Conference on Vibratory Pile Driving Transvib 2002.

Table 6. Soil description, penetration speed and skin friction

| depth below ground surface | soil description | penetration speed | | skin fric- |
| | | mean | max | tion |
| --- | --- | --- | --- | --- |
| 0 to 8 m | coarse sand | 0.4 m/min | 1.2 m/min | 45 kPa |
| 8 to 14 m | clayey sand | 2 m/min | 6 m/min | 20 kPa |
| 14 to 19 m | alternation of sand and clay | 0.4 m/min | 1.2 m/min | 35 kPa |

## ACKNOWLEDGEMENT

The tests reported in this paper were carried out in the framework of the French national project on vibratory driving. This project is managed by IREX and forms part of the operations of the RGC&U network. The French Ministry of Public Works and the harbor Authority of Nantes Saint-Nazaire are acknowledged for their financial and technical support.

The author is grateful to Jean-François Vanden Berghe for his help in the analysis of vibratory driving records. The spectral analyses were carried out thanks to Noël Huybrechts. Jean-François Semblat and Alain Holeyman are acknowledged for their valuable observations during the paper review.

## REFERENCES

Borel S., Bustamante M., Gianeselli L. 2002. A comparative field study of the bearing capacity of vibratory and impact driven sheet piles. *Proceedings international conference on vibratory pile driving and deep soil compaction, Louvain-la-Neuve.*

Briaud J.-L., Coyle H.M., Tucker L.M. 1990. Axial response of three vibratory and three impact driven H piles in sand. *Transportation Research Record. No. 1277. pp. 136-147.*

Bustamante, M. & Doix, B. 1991. A new model of LPC removable extensometer. *Proceedings 4th international conference on piling and deep foundation, Stresa, 7-12 April 1991.* Rotterdam: Balkema.

Bustamante, M., Borel, S., Gianeselli, L. 1999 Essais de chargement statique sur pieux instruments – pieu métallique fermé battu – pieu foré tarière creuse. *Rapport d'essai LPC, 9 pages + Appendix.*

Franke E., Mazurkiewicz B. 1987. The influence of the pile installation method on its bearing capacity. *Proceedings, 8th Polish national conference on soil mechanics and foundation engineering. vol. 2, pp. 533-538.*

Mazurkiewicz B. 1975. The influence of vibration of piles on their bearing capacity. *Proceedings, 1st Baltic Conference on Soil Mechanics and foundation engineering, Gdansk, section 3, pp. 144-153.*

Mosher R.L. 1990. Axial capacity of vibratory-driven piles versus impact driven piles. *Transportation Research Record. No. 1277. pp. 128-135.*

Règles techniques de conception et de calcul des fondations des ouvrages de génie civil. *Cahier des clauses techniques générales applicables aux marchés publics de travaux. Fascicule 62 – Titre V 1993.* Ministère de l'équipement, du logement et des transports. Paris: Textes officiels (in French).

# International prediction event of vibratory pile driving

J-F. Vanden Berghe
*THALES GeoSolutions, Brussels, Belgium*

ABSTRACT: A vibratory piles driving prediction was organised within the framework of the TransVib 2002 conference in collaboration with the IREX project, "Projet National Vibrofonçage". Instrumented piles were vibrodriven on the site of Montoir in France in August 2001. Ten class A predictions were submitted before the performance of the test. The present paper presents the organisation of the prediction event and describes briefly the models used by the predictors. The results of these predictions are also presented and compared with the measurements.

## 1 INTRODUCTION

Within the framework of the French IREX-project "Vibrofonçage", instrumented tubular piles were installed using vibratory driving technique on the site of Montoir, which belongs to the Port de Nantes in France. After installation, one of the vibrodriven piles was statically load tested.

Based on these tests, the organising committee of the TransVib2002 conference in collaboration with the IREX project took the opportunity to organise a class A international prediction event. Prior to the tests, a project synopsis had been prepared to invite interested parties to make predictions. It included a description of the piles, the characteristics of the vibrator, the available information regarding the site and the format of the prediction. 10 predictions were submitted before the performance of the tests.

The present papers focuses on the organisation of the prediction event, the models used by the predictors and the comparison between the predictions and the tests results. The site investigation, the tests organisation and the discussion of the measurements are described in a companion paper (Borel & al., 2002) and are only summarised in the following paragraphs.

## 2 SOIL INVESTIGATION

The geotechnical investigation of the site of Montoir was conducted in 1999 when an other campaign of static loading tests was performed on impact driven piles (LCPC, 1999). The geotechnical survey of the site was performed based on cone penetration tests (CPT), SPT profile, PMT pressuremeter and laboratory tests. The results of these tests are described in Borel (2002) and Simecsol (1999). The analysis of these tests indicated to the soil stratigraphy presented in Table 1.

Table 1: Soil Stratigraphy in Montoir (Borel,2002)

| depth below ground surface (m) | Soil description |
|---|---|
| 0 to 4.5 | brown coarse sand with gravel |
| 4.5 to 10 | grey-black coarse grained sand |
| 10 to 13 | slightly clayey coarse to fine sand |
| 13 to 22.5 | alternating with coarse to fine sand layers (40-100 cm thick) and grey-black sandy clay layers (10-30 cm thick) |
| 22.5 to 36 | finely interbedded (1 to 20 cm) alternating with fine sand and mud clay (locally named "jalle") |

# 3  PILE AND VIBRATOR CHARACTERISTICS

## 3.1  Pile dimensions

For the prediction exercise, the tubular pile was assumed to have an external diameter of 339.7 mm and a wall thickness of 14.2 mm. The pile length was taken equal to 32 m. The pile base was assumed closed.

However, during the execution of the tests, steel profiles were welded along the pile in order to protect the instruments and the cables. The actual size of the pile during the test was then slightly different from the dimensions assumed for the prediction. The difference between the assumed and the actual dimensions is reported in Table 2.

Table 2: Comparison between assumed and actual pile dimensions

|  | Assumed values | Actual values |
|---|---|---|
| Steel Section | 145cm² | 160cm² |
| Perimeter | 1.07m | about 1.40m |
| Total Length | 32m | 32m |
| Toe Section | 906cm² | 1130cm² |
| Total pile weight | 3650kg | about 3700kg |

## 3.2  Vibrator characteristics

An ICE 815 vibrator was chosen for the vibratory driving of the tubular piles. The characteristics of the vibrator, its clamps and the power pack are summarised in Table 3.

Table 3: Assumed characteristics of vibrator, clamps and power pack

| VIBRATOR | |
|---|---|
| Vibrator Type | ICE 815 |
| Eccentric moment | 46 kg.m |
| Frequency Range | 800/1600 rpm |
| Nominal centrifugal force | 1250 kN |
| Maximum Amplitude, excl. Clamps | 26 mm |
| Maximum static traction force | 400 kN |
| Hydraulic power kW/HP | 300 kW/405 HP |
| Maximum working pressure | 340 bar |
| Maximum flow | 570 l/min |
| Vibrating Mass,excl. Clamp | 3550 kg |
| Static Mass | 2450 kg |
| CLAMP AND HYDRAULIC HOSES | |
| Clamp type | 100 TP |
| Clamp Mass | 2000 kg |
| Maximum Clamp force | 1000 kN |
| Mass hydraulic hoses | 750 kg |
| | |
| Total Vibrating mass, incl. Clamps | 6050 kg |
| Total Mass, incl. Clamps | 8500 kg |
| POWER PACK | |
| Power pack type | type 500 Caterpillar 3306 DITA |
| Power | 384 kW / 522 HP |
| Maximum regime | 2300 rpm |
| Maximum pressure | 340 bars |
| Mass | 7100 kg |

# 4  PREDICTION ORGANISATION

Two approaches were proposed for the predictions: one based on a pragmatic experience and engineering judgement (results oriented) and the second based on a theoretical approach (process oriented).

For the pragmatic prediction, the following questions were asked:
- Will the vibrator drive the pile to the desired depth?
- What will be the maximum vibration level observed around the pile?
- What would be the refusal depth reached with that vibrator (assuming a pile long enough)?

For the theoretical prediction, predictors were asked to provide solutions for 3 frequencies of the vibrator - 800, 1200 and 1600rpm - (in order to cover the range of frequencies that might be used by the operator) and for each meter of penetration:
- the average penetration speed of the pile,
- the amplitude of the displacement at the top of the pile,
- the acceleration at the top and the bottom of the pile (peak downward, peak upward and RMS),
- the axial strain at the top and the bottom of the pile (peak downward, peak upward and RMS),
- the peak particle velocity of ground surface at a radius of 6.4 m, 12.8m and 19.2m from the pile.

For the 3 frequencies of the vibrator (800rpm, 1200rpm and 1600rpm) and for a penetration of 10m, 20m, and 30m, predictors were also to provide the variation over one period of the vibrator of:
- the acceleration,
- the velocity,
- the displacement,
- the total soil resistance,
- the shaft resistance,
- the toe resistance,
- the particles velocity of the ground surface at a radius of 6.4 m, 12.8m and 19.2m m from the pile.

The predictors had the opportunity to request that their prediction remains anonymous.

The call for predictions was sent the 28th March 2001. The documents available were sent as soon as the registration form was received. The deadline was fixed for the 15th July 2001. The vibratory tests were performed the 31st august 2001. In May 2002, the tests results were sent to the predictors in order to provide them the opportunity to present their prediction in more detail, to discuss their results regarding the tests results and to be able to perform a post-event reanalysis.

Table 4: Presentation of the predictions

| Prediction | Predictor | Affiliation | Country | Type of model |
|---|---|---|---|---|
| Prediction 1 | L. Aaltonen | Fundus OY | Finland | Longitudinal 1D model |
| Prediction 2 | R. Cudmani | Univesity of Karlsruhe | Germany | Longitudinal 1D model |
| Prediction 3 | D Guillaume | ICE France | France | Longitudinal 1D model |
| Prediction 4 | A. Holeyman | Université catholique de Louvain | Belgium | Radial 1D rigid model |
| Prediction 5 | A. Holeyman | EarthSpectives | USA | Semi emperical equation |
| Prediction 6 | G. Jonker | IHC Hydrohammer | The Netherlands | Semi emperical equation |
| Prediction 7 | P. Middendorp | TNO Profound | The Netherlands | Longitudinal 1D model |
| Prediction 8 | J-G Sieffert | Ensais | France | Rigid model |
| Prediction 9 | R. Van Foeken | IHC Foundation Equipment | The Netherlands | Longitudinal 1D model |
| Prediction 10 | *anonymous* | *anonymous* | | Finite element analysis |

## 5 PREDICTION PRESENTATION

After a general presentation of the predictions, the following paragraphs describe briefly the models used by the different predictors.

### 5.1 General

10 predictions were submitted before the execution of the tests. Table 4 presents the label used for each prediction, the name of the predictor and his affiliation.

The type of model that was used for each prediction is also indicated in Table 4. All the predictions were founded on a theoretical approach. Five predictions used a 1D longitudinal model adapted from the Smith's classic lumped parameter model. Two predictions were based on a semi-empirical equation deduced from force equilibrium during vibratory driving. The two last predictions integrated the equation of motion to calculate the penetration speed of the pile assuming the pile as a rigid body.

Table 5 indicates the results that were provided by each predictor.

### 5.2 Prediction 1

Prediction 1 used a 1D longitudinal model adapted from the Smith's classic lumped parameter model (Fig 1). The pile is divided in a series of elements that are interconnected with springs which stiffness depends on the pile characteristics. The first element represents the static mass and is connected by a soft spring, whereas the second element simulates the exciter block and is subjected to a sinusoidal force. Along the pile shaft, the soil behaviour is modelled using a spring-slider-dashpot system, i.e. a perfect elasto-plastic behaviour combined with a dynamic resistance. At the pile toe, the same type of dynamic system is used but without allowing nega-

Table 5: Provided results of each prediction

| | Provided results | | | | | | | | | | | | |
|---|---|---|---|---|---|---|---|---|---|---|---|---|---|
| | | As a function of the the pile penetration | | | | | As a function of the time at 10m, 20m and 30m | | | | | | |
| | Nbr of Frequecies | Pile Amplitude | Acceleration | Penetration Speed | Pile Deformation | Vibration Level | Acceleration | Velocity | Pile displacement | Total Pile Resistance | Shaft Resistance | Toe Resistance | Vibration Level |
| Prediction 1 | 3 | X | X | X | X | | X | X | X | X | X | X | |
| Prediction 2 | 3 | X | X | X | X | | X | X | X | X | X | X | |
| Prediction 3 | 1 | X | | X | | | | | | | | | |
| Prediction 4 | 3 | X | | X | | X | (X) | (X) | (X) | (X) | (X) | | |
| Prediction 5 | 3 | X | X | X | | | | | | | | | |
| Prediction 6 | 3 | X | X | refusal depth | | | | | | | | | |
| Prediction 7 | 3 | X | X | X | | | (X) | (X) | (X) | | | | |
| Prediction 8 | 3 | X | X | X | | | X | X | X | X | X | X | |
| Prediction 9 | 3 | X | | X | | | | | | | | | |
| Prediction 10 | 1 | | | | | X | | | | | | | |

The crosses between brackets indicate that the results were submitted in format incompatible with the data processing of the prediction. These results are not presented in this paper.

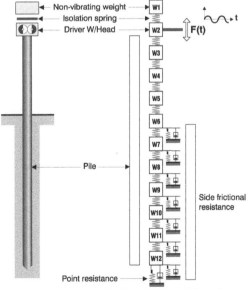

Fig 1: 1D longitudinal model adapted from the Smith's classic lumped parameter model

tive resistance. The input parameters of the model are: the maximum resistance of each element, the quake, the damping factor, the length of each element, the area of the pile section and Young's modulus of the pile.

The predictor deduced the soil parameters based on the friction angle provided in the report of the geotechnical investigation.

## 5.3 Prediction 2

Prediction 2 was performed using the model described in Cudmani (2002). This model computes the pile displacements by integrating the one-dimensional stress wave equation in an implicit finite element scheme (Fig 2). The soil reaction along the pile shaft is divided in 2 terms: the shaft resistance and the dynamic damping. The shaft resistance is modelled by an elasto-plastic hysteresis loop depending of the pile displacement whereas the dynamic damping is calculated using a simplified equation of wave damping in an elastic medium. The soil resistance at the pile toe considers different phases depending of the pile toe position relative to the soil and a cavitations phase is introduced when the pile toe is not touching the soil. Since this behaviour is rate independent, additional dynamic damping is added at the pile toe. This dynamic damping is based on the analytical equation of the wave damping in a semi-infinite medium. This model depends on 9 parameters: 4 for the toe resistance, 3 for the shaft resistance and 2 for the dynamic damping. The method

to evaluate these parameters is described in Cudmani (2002).

## 5.4 Prediction 3

The calculations of prediction 3 were performed with the TNOWAVE programme for impact and vibratory hammers combined with the TNOVPD module for simulation of vibratory pile driving. TNOWAVE is a one-dimensional stress wave simulation program to simulate static and dynamic effects in piles. TNOWAVE models the load introduced to the pile top, the foundation pile, and soil behaviour at pile shaft and pile toe. The pile is modelled with the Smith's classic lumped parameter model (Fig 1). The pile is divided in a series of elements that are interconnected with springs which stiffness depends on the pile characteristics. The shaft and toe resistance are modelled with a spring-slider-dashpot system. The input parameters of the soil model of this software are the evolution of the maximum resistance, the quake and the damping factor along the pile and at the pile toe.

These input parameters were deduced from the results of the SPT test performed during the site investigation.

## 5.5 Prediction 4

The model used in prediction 4 computes the pile penetration and the wave propagation around the pile by modelling the soil with a set of concentric rings

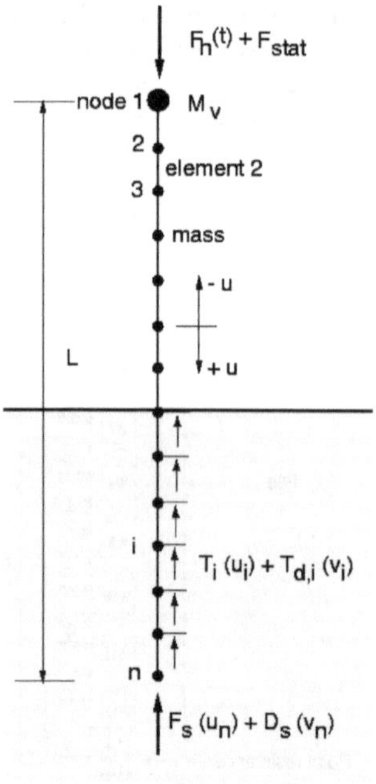

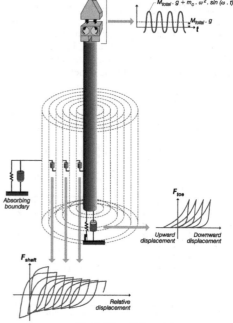

Fig 2: Finite element scheme considered in prediction 2

Fig 3: 1D radial model

196

possessing individual mass (Fig 3) and transmitting forces to their neighbouring ones (Holeyman – 1993a). The pile and the soil elements are assumed to be rigid bodies. To calculate the displacement of the pile as well as the wave propagation around the vibrating profile, the model numerically integrates the equation of motion of each cylinder based on their dynamic shear equilibrium in the vertical direction. The model calculates the shear stress along the pile shaft and between soil elements by constructing the hysteretis loops as a function of the displacements of the pile and soil elements. The model considers also the decrease of the soil resistance and the soil liquefaction induced during cyclic loading. The base resistance is modelled with a spring-slider-dashpot system with zero toe resistance when the pile is moving upward.

The input parameters of the model (HYPERVIB II) used for this analysis were deduced from the CPT tests.

## 5.6 Prediction 5

Prediction 5 used a semi-empirical equation (Holeyman – 1993b) allowing the evaluation of the average penetration speed based on the peak acceleration of the free hanging vibrator-pile system. This equation assumes the soil behaviour is perfectly plastic, and the net acceleration during a cycle can be deduced from the difference between the sinusoidal driving force and the opposing soil resistance. The equation considers also the difference of behaviour between the base and toe resistance. The average penetration is calculated by intuitively integrating the net downward and upward acceleration over a complete cycle. The soil resistance is evaluated based on the results of CPT tests, reduced as a function of the friction ratio and the acceleration to take into account the degradation of the soil strength. This model has been verified and calibrated based on different results of full-scale tests (Holeyman, 1996).

The predictor evaluated the input parameters of this model (HYPERVIB I) based on the results of the CPT test.

## 5.7 Prediction 6

The model of prediction 6 is based on force equilibrium and determines whether or not the vibrator can overcome the resisting forces applied to the pile during the vibratory driving. This model does not provide an estimation of the penetration speed. It only verifies whether the maximum driving force applied by the vibrator is greater than the resistance of the soil.

The β model (Jonker - 1987) calculates for each penetration depth the static and dynamic soil resistance, the remaining amplitude at the pile toe and the acceleration at the toe. The model deduces from these values the so-called beta parameter defined as the ratio between the dynamic and static forces. The obtained beta parameter is compared with limit values deduced from experience in different types of soil. If the parameter is lower than the limit value, the model considers that the vibrator is able to drive the pile into the soil at the considered depth. The background of this model and Prediction 6 are described in more details in Jonker (2002)

## 5.8 Prediction 7

Prediction 7 were performed with the TNOWAVE programme. This programme was described in the paragraph concerning prediction 3.

## 5.9 Prediction 8

Prediction 8 was performed using the programs VIBPIEU and BRAXUUS (Sieffert, 2002).

VIBPIEU calculates the displacement amplitude of the vibrating system in the free field (without soil influence). This calculation is performed for 2 cases: assuming the pile as a one dimensional continuum medium or as a rigid body. In both cases, it calculates the amplitude of the vibrating part of the vibrator and the amplitude at the pile extremities.

BRAXUUS calculates the pile displacement during the vibratory driving by integrating the equation of motion. The model considers the pile as a rigid body and calculates the soil resistance assuming a perfect plastic behaviour. The shaft resistance is directly mobilised in the opposite direction of the pile displacement whereas the toe resistance is directly mobilised only when the pile penetrate deeper than the maximum depth reached during the previous cycles. A detailed description of this model is available in Sieffert (2000 and 2002)

## 5.10 Prediction 9

The calculations for prediction 9 were performed with the TNOWAVE programme for impact and vibratory hammers. This programme was described in paragraph 5.3.

## 5.11 Prediction 10

The model of prediction 10 used the axisymmetrical finite element model DIANA to simulate the vibration induced around pile during vibratory driving. The pile and the pile-soil interaction were modelled by the program VP-damwand developed by TNO Bouw. This program was created based on the assumptions and models of pile-soil interaction during vibratory driving proposed by Dierssen (1994).

The input parameters of this prediction were deduced from the CPT test performed during the geotechnical site investigation.

# 6 PREDICTION RESULTS

## 6.1 Test results

The results of the vibratory driving test performed on the site of Montoir are presented in details in Borel (2002). The following paragraph summarises the main results and highlights the differences between the parameters assumed for the prediction and the actual values of these parameters.

Fig 4-a shows the evolution of the penetration speed of one of the piles driven in Montoir. The evolution of the penetration speed can be characterised as follow:

- 0.4m/min between 0 and 7m,
- 6m/min between 7m and 13m,
- 0.4m/min between 13m and 18m
- Refusal at 18m (i.e. speed < 0.1m/min)

The frequency of the vibrator was about 1350rpm and decrease slightly from 1380rpm to 1300rpm during the driving of the pile (Fig 4-d).

The maximum upward and downward accelerations measured at the pile toe and the pile head are shown on Fig 4-b. The downward accelerations (maximum value) are almost equal whereas the upward acceleration (minimum value) is higher at the pile toe than the pile head.

The measured accelerations were integrated to calculate the value of the displacement amplitudes at the pile toe and the pile head (Fig 4-c). According the vibrating mass and the eccentric moment, the theoretical free hanging amplitude of the pile displacement is equal to 10.7mm peak to peak. The amplitudes deduced from the test are almost constant during the penetration and remain close to the theoretical free hanging amplitude. On the other hand, the amplitudes observed at the pile toe are a somewhat higher than at the pile head and decrease progressively from 15mm to 13mm during the pile penetration.

## 6.2 Differences between assumed parameters and actual parameters

During the execution of the test, the following modifications of the parameters proposed for the prediction were made:

- The cross section of the pile base was increased for the protection of the transducers and cables.
- The frequency of the vibrator was about 1350rpm.
- The vibrating mass of the pile-vibrator system was 8550kg instead of 9650kg.
- The geophones to measure the vibrations induced around the pile were placed at a distance of 6m 12m and 18m instead of 6.4 m, 12.8m and 19.2m.

## 6.3 Comparison between predictions and tests results

### 6.3.1 Penetration speed

Fig 5 compares the predictions of the penetration speed performed for the 3 requested frequencies (800rpm, 1200rpm and 1600rpm). Some predictors preferred to perform the prediction for the highest frequency with a value slightly lower than 1600rpm. These predictions are plotted together with the predictions with a frequency of 1600rpm.

The test results (frequency = 1350rpm) are plotted above the predictions with a frequency of 1200Hz and 1600Hz.

### 6.3.2 Displacement amplitude

The predicted amplitudes of the pile displacement are plotted on Fig 6 for the 3 considered frequencies. The amplitudes deduced from the measurements at the pile toe and the pile head are also shown on the figure.

The majority of the predictions indicates vibration amplitudes between 10mm and 15mm and fits quite well with the measured data.

### 6.3.3 Pile acceleration

The minimum and maximum acceleration during each cycle is drawn versus depth on Fig 7. The acceleration is defined positive in the downward direction.

The predicted accelerations are in the same range and correspond quite well with the measured values.

Figs 8 and 9 show the evolution of the predicted acceleration during 2 periods of the vibrator for penetration depths of 10m, 20m and 30m and for a frequency equal respectively to 1200rpm and 1600rpm. The results of the test measured at 10m depth are also plotted on the predictions. The calculated results generally fit well with the measurements. Some predictions seem also to be able to simulate the different phases of the curve.

### 6.3.4 Soil resistance

On Figs 10 to 12, the evolution of the predicted total soil resistance, the toe resistance and the shaft resistance during 2 cycles of vibrator are plotted at 3 different penetration depths (10m, 20m and 30m). These figures indicate the evolution as a function of the time (left column) and as a function of the pile displacement (right column).

### 6.3.5 Induced vibration around the pile

Only 2 predictors submitted an evaluation of the vibrations induced around the pile. These predictions are compared with the experimental data on Fig 13. As explained in paragraph 6.2, the position of the geophone was slightly different than assumed for the prediction (6m, 12m and 18m instead of 6.4 m,

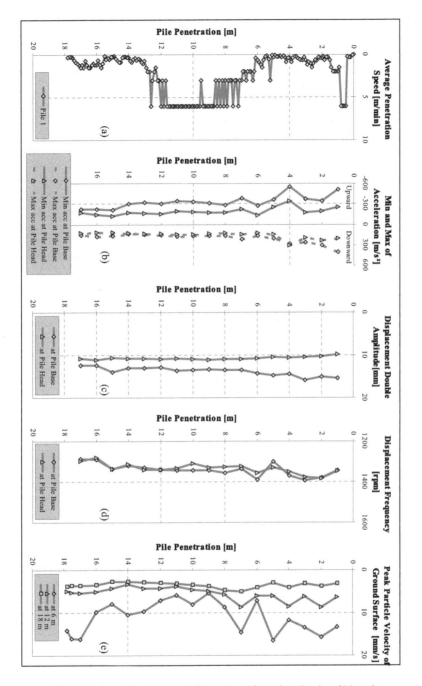

Fig 4: Results of the vibratory driving test performed on the site of Montoir.

199

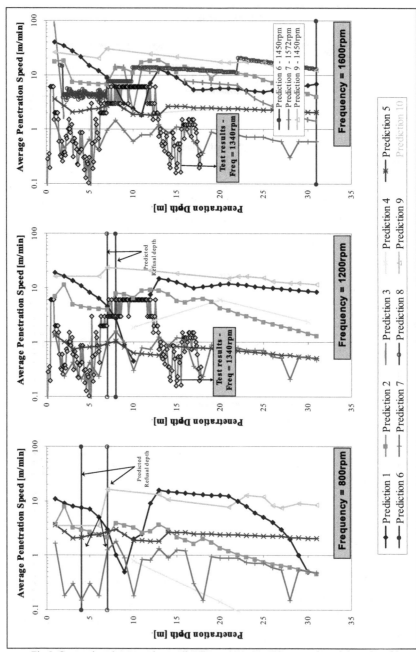

Fig 5: Comparison between the predicted average penetration speeds and the test results
for 3 different frequencies (800rpm, 1200rpm and 1600rpm).

200

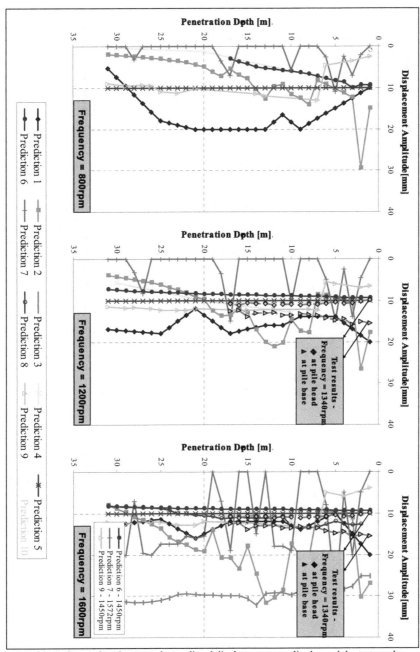

Fig 6: Comparison between the predicted displacement amplitudes and the test results for 3 different frequencies (800rpm, 1200rpm and 1600rpm).

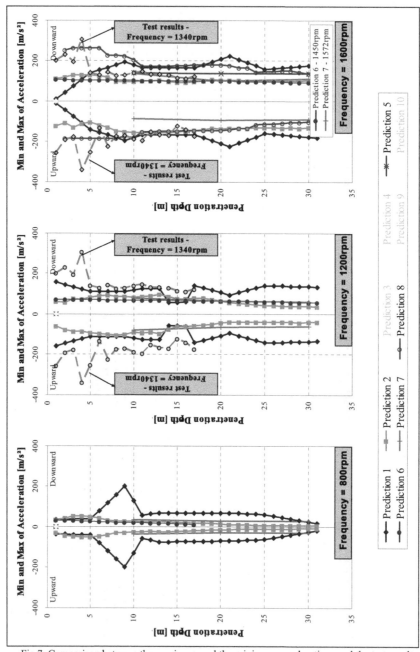

Fig 7: Comparison between the maximum and the minimum accelerations and the test results for 3 different frequencies (800rpm, 1200rpm and 1600rpm).

202

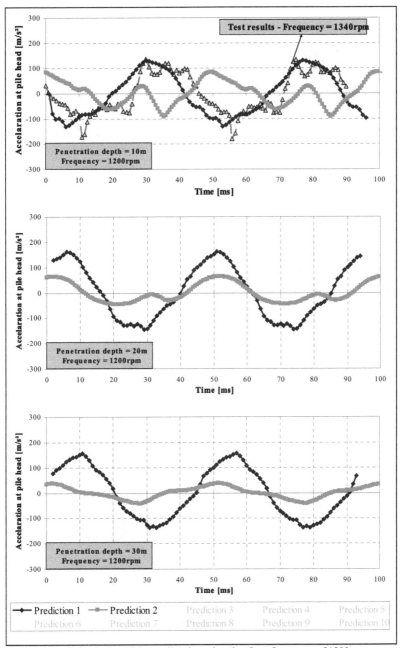

Fig 8: Evolution of the predicted acceleration for a frequency of 1200rpm
for a penetration depth of (a) 10m, (b) 20m and (c) 30m.

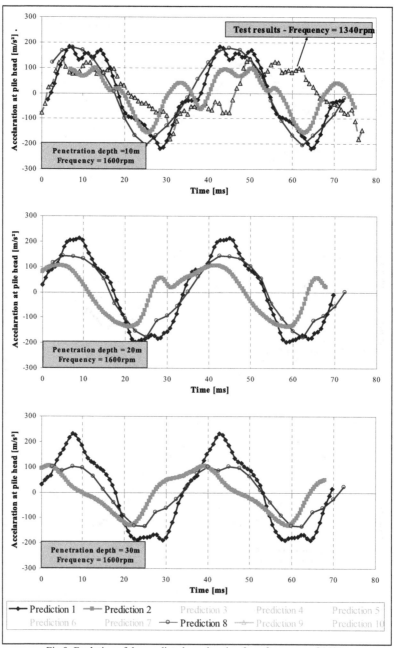

Fig 9: Evolution of the predicted acceleration for a frequency of 1600rpm
for a penetration depth of (a) 10m, (b) 20m and (c) 30m.

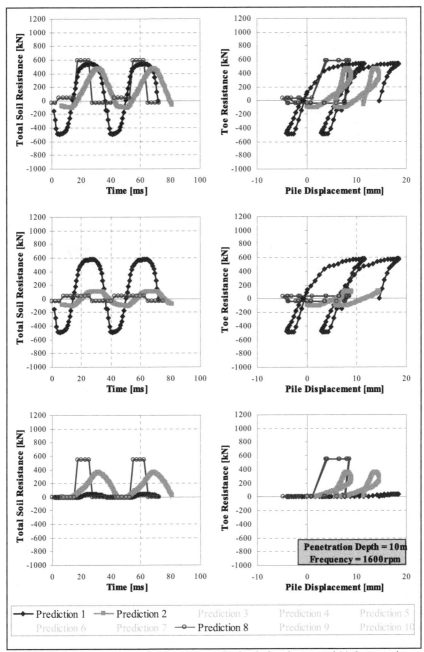

Fig 10: Evolution of (a) the total soil resistance, (b) the shaft resistance and (c) the toe resistance for a penetration depth of 10m and a vibrator frequency equal to 1600rpm.

205

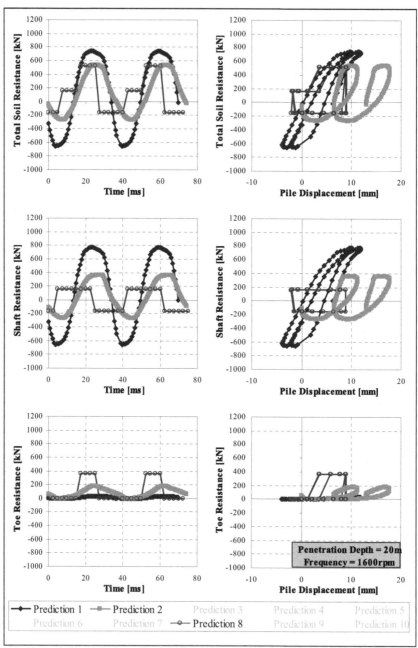

Fig 11: Evolution of (a) the total soil resistance, (b) the shaft resistance and (c) the toe resistance for a penetration depth of 20m and a vibrator frequency equal to 1600rpm.

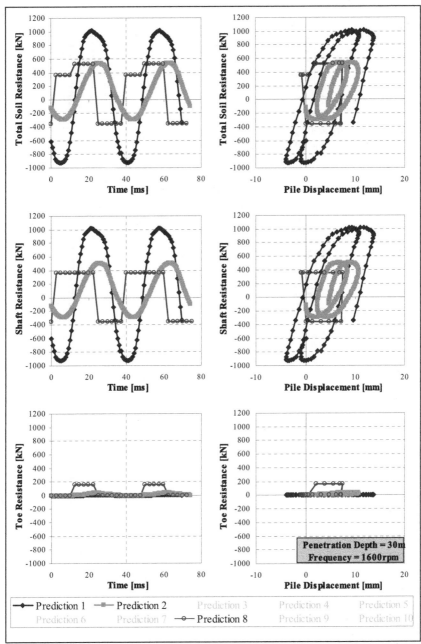

Fig 12: Evolution of (a) the total soil resistance, (b) the shaft resistance and (c) the toe resistance for a penetration depth of 30m and a vibrator frequency equal to 1600rpm.

207

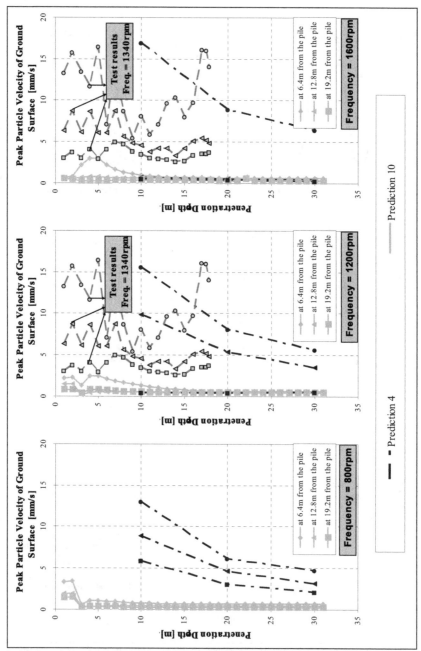

Fig 13: Comparison between the predicted peak particle velocity around the pile and the test results for 3 different frequencies (800rpm, 1200rpm and 1600rpm).

12.8m and 19.2m). However, this difference should not change significantly the results.

The analysis of the results shows a relatively large difference between the measured and the calculated values on the site of Montoir.

## 7 CONCLUSIONS

The present paper has presented the results of the prediction event of vibratory pile driving that were organised in collaboration with the French Irex project. 10 predictors participated in this exercise providing an overview of some of the available models to predict the vibrator performances during vibratory driving.

## 8 ACKNOWLEDGEMENT

The author would like to thank the members of the French Irex project "projet national de vibrofoncage" who available the information from the site investigation and the results of the tests to perform this prediction event. The author gratefully acknowledges Mr. S. Borel for his help in the analysis of the tests results. This exercise would not have been possible without the active participation of the predictors that bravely took up the challenge. The organisers thank them for their contributions. Special thanks are going to David Cathie and Alain Holeyman for their help and their comments during the writing of this paper.

## 9 REFERENCES

BOREL S, GIANESELLI L., DUROT D., VAILLANT P., BARBOT L., MARSSET B.&. LIJOUR P. (2002), Full-scale behaviour of vibratory driven piles in Montoir, Proceedings of the TranVib2002 conference, Louvain-la-Neuve, 11p.

CUDMANI R. O., HUBER G., GUDEHUS G.. (2002), A mechanical model for the investigation of the vibro-drivabilty of piles in cohesionless soils, Proceedings of the TranVib2002 conference, Louvain-la-Neuve, 11p.

CUDMANI R. O. (2002), Class A prediction of the vibrodrivability of steel pile in Montoir, Proceedings of the TranVib2002 conference, Louvain-la-Neuve.

DIERSSEN G. (1994), Ein bodenmechanisches Modell zur Beschreibung des Vibrationsrammens in körnigen Böden, Karlsruhe.

HOLEYMAN, A. (1993a) "HYPERVIBIIa, An detailed numerical model proposed for Future Computer Implementation to evaluate the penetration speed of vibratory driven sheet Piles", Research report prepared for BBRI, September, 54p.

HOLEYMAN, A. (1993b), "HYPERVIB1, An analytical model-based computer program to evaluate the penetration speed of vibratory driven sheet Piles", Research report prepared for BBRI, 23p.

HOLEYMAN, A., LEGRAND, C., AND VAN ROMPAEY, D., (1996). "*A method to predict the driveability of vibratory driven piles*". proceedings of the 3rd international conference on the application of stress-wave theory to piles, pp 1101-1112, orlando, u.s.a., 1996.

JONKER, G., (1987). "Vibratory Pile Driving Hammers for Oil Installation and Soil Improvement Projects". Proceedings of Nineteenth Annual Offshore Technology Conference, Dallas, Texas, OTC 5422, pp. 549-560.

JONKER, G. (2002), Vibrateability Predictions using the $\beta$-Method, Proceedings of the TranVib2002 conference, Louvain-la-Neuve, 7p.

LCPC (1999), Port de Montoir T.M.D.C. 4 – Essais de chargement – Rapport technique

SIEFFERT, J.-G. (2000). Prévision de la pénétrabilité. Méthode ENSAIS – Logiciel BRAXUUS. Research report: IREX : 29 pages

SIEFFERT, J.-G. (2002), Vibratory pile driving analysis: a simplified model, Proceedings of the TranVib2002 conference, Louvain-la-Neuve, 8p.

SIMECSOL (1999), Port de Montoir T.M.D.C. 4 – Campagne Geotechnique et instrumentation en cours de battage d'un pieu d'essai – Note de presentation.

*Vibratory prediction event*

*Individual contributions*

# Vibrateability Predictions using the ß-Method

G. Jonker
*IHC Hydrohammer BV, Kinderdijk, The Netherlands*

N. van der Zouw
*New Institute, Gouda, Netherlands*

ABSTRACT: As part of an extensive test pile programme on the soil characteristics which determine to a great extend the vibrateability of piles, a prediction event was organised prior to pile installation. Purpose was to validate the existing prediction methods on their accuracy, as well as to determine the sensitivity of some of the parameters.
The author participated in the contest using the β-method as developed by ICE BV some 15 years ago.
The paper describes the background of the method, its input and output parameters, and the prediction and postdiction results for the test piles. Test piles were 339x14mm closed ended, 32m long tubular piles, driven by an ICE-815 vibratory hammer.

## 1 INTRODUCTION – BACKGROUND

In the late 1950-s the use of vibratory hammers became increasingly popular, especially because of their high speed of installation, low weight and ease of operation. Ever since that time many scientific and research institutes, individuals and companies have developed methods to predict the pile driving performance by vibration, the so-called 'vibrateability' of piles. A summary of these methods is given in reference 1 (D.C. Warrington, 1990). All methods can be divided into 4 main groups, the Parametric Methods, the Energy Methods; Methods based on Laboratory and Model Tests, and Wave Equation Methods.

The author developed the β-method, a Parametric Method, during his engagement with manufacturer ICE BV in the Netherlands in the mid 1980's. Since that time the method and simulation program has been continuously validated by back analyses of field observations. Although it is recognized already for a long time that the wave-equation method is the best tool to simulate pile-driving behavior (Jonker and Middendorp, 1988). This method however was lacking sufficient calibration and validation in its early days to be used for prediction purposes. In the last couple of years tremendous efforts and progress have been made in this respect, this being one of the reasons that ICE is presently transferring their extensive experience with the β-method program into Profound's VDPWAVE wave equation program. The extensive testpile program in Montoir will highly contribute to further validate every existing method but especially the wave-equation method.

## 2 VIBRATORY PILE DRIVING

The operating principle of vibratory hammers is supposed to be known to the reader and is no further explained in this paper.

Main characteristics which determine the capacity of vibratory hammers are Eccentric Moment ($M_e$ in kgm), operating Frequency ($\omega$ in rad/s, or f in cycles per minute) and Centrifugal Force ($F_c$ in N).
The relation between these characteristics is given by:

$$F_c = M_e \cdot \omega^2 \qquad (1)$$

Vibratory hammers are normally classified in 'Low Frequency' and 'High Frequency' hammers, the interface between the two is arbitrarily taken by the author as 30 Hz (f=1800 cpm). This frequency is considered to be sufficiently above the natural frequency of most soils (10-20 Hz, Jonker, 1987) to avoid damages to buildings, or to cause severe settlements of soils due to resonance effects of the generated vibrations.

Contrary to impact driving of piles, whereby most of the static capacity of the pile has to be overcome by each blow, the soil loses most of it's static resistance when subjected to vibrations. This is the main reason why only a relative low (centrifugal) force can drive a (low displacement) pile to relative great depths. The vibrations of the pile are transferred to the adjacent soil and, in granular soils, results in a kind of random displacement (movement) of the individual grains. The contact (force) between the grains is almost or completely lost, reducing the internal friction to almost zero. The loss of internal friction results automatically in a loss of shaft-friction between pile and soil. It further facilitates the displacement of soil underneath the toe of the pile.

In clay type soil the vibrations and up-and-down displacement of the pile take care for soil-remolding, building up of excess water-pressure and for soil fatigue. It results in a (much) lower shaft-friction compared to the situation whereby the pile is statically loaded.

When vibratory driving is stopped, the soil regains generally most of its temporary loss of resistance. The effects of the vibrations depends on the original density of the material, grainsize distribution, water content, frequency and amplitude of vibrations, etc. Loose sands are normally compacted after the vibrations; very dense soils are loosened up.

## 3  THE β-METHOD FOR VIBRATORY PREDICTIONS.

As stated above, the β-method belongs to the group of Parametric Prediction Methods. The basic principle of the method is based on an equilibrium of forces (fig. 1), i.e. the equilibrium between the centrifugal force of the vibratory hammer ($F_c$) and the total vibratory resistance of the soil. Jonker (1987) introduced the term SRV (Soil Resistance

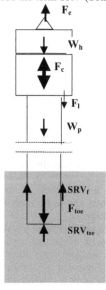

Figure 1. Forces acting on pile

during Vibratory driving) in accordance with the abbreviation SRD (Soil Resistance to Driving) for impact hammers.

To achieve penetration or extraction, the 'driving forces' should be larger than the 'reaction forces'.

In formula form for driving:

$$F_c + F_1 + W_h + W_p > SRV_f + > SRV_{toe} \qquad (2)$$

And for extraction:

$$F_c + F_e > W_h + W_p + SRV_f \qquad (3)$$

Whereby:

| | |
|---|---|
| $F_c$ | : centrifugal force |
| $F_1$ | : down trust (line pull) |
| $F_e$ | : static line pull on extraction head |
| $W_h$ | : total weight of hammer incl clamp |
| $W_p$ | : weight of pile |
| $SRV_f$ | : shaft resistance to vibratory driving |
| $SRV_{toe}$ | : toe resistance to vibratory driving |

Similarly to pile driveability predictions for impact hammers is the calculation of the SRV values based on the static capacity of the pile. For an unplugging pile behavior, which is nearly always the case in vibratory driving operations, the static capacity is calculated by either:

$$Q_t = F_o + F_i + Q_w \quad \text{or} \quad Q_t = F_o + Q_w \qquad (4)$$

Whereby:

| | |
|---|---|
| $Q_t$ | : total static capacity |
| $F_o$ | : total outside shaft friction |
| $F_i$ | : total inside shaft friction (open pipe piles) |
| $Q_w$ | : total endbearing under pile wall |

If β-factors are introduced to determine the ratio between the vibratory driving resistance and the static resistance, the formula is transformed to:

$$SRV_t = \beta_o F_o + \beta_i F_i + \beta_t Q_w \qquad (5)$$

Apart from being a function of the usual soil parameters, one should keep in mind that the β-values are different for low frequency and high frequency hammers, for open and closed ended piles, and may differ for piling and extraction operations. The β-values depends further on the acceleration of the pile (figure 2) and may depend on the amplitude of motion also.

Low $\beta_{min}$-values (at high accelerations) are achieved in loose sandy soils and very soft clayey soils, and may be as low as 0,05. For hard over-consolidated clays the β-values may be as high as 0,05. The $\beta_{min}$-values for the tip may vary from 0,1 to 0,8 depending on the density of the soil.

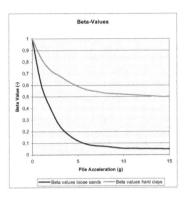

Figure 2. β--values as function of acceleration. (Japanese publication, source unknown)

## 4 SIMULATION PROGRAM

In 1987/1988 ICE BV in The Netherlands has put all above formulas and graphs into a MS-DOS based computer simulation program (Jonker 1987, Ligtering, Van Zandwijk, Middendorp 1990, Ligterink and Martin 1992) In this program, the vibrateability of a pile is calculated in a iterative step-by-step procedure. Every 1 meter penetration the equilibrium as mentioned in paragraph 2 is calculated taking into account the modification of the β-factor as a function of the acceleration at the pile toe. For the calculation of the acceleration and displacement-amplitudes the hammer (dynamic weight, including clamp) and the pile can be treated as one single mass. This is considered acceptable as vibratory driven piles are generally not longer than 10-20m, i.e. the stress-wave travelling time from head to toe is not more 2-4 ms. The duration of a half sine-wave of a 20 Hz and 40 Hz frequency hammer last 25 respectively 12,4 ms. One may with sufficient accuracy assume that the pile is 'equally' loaded and that the differences between the force at the head and the force at the toe are minimal. For the same reason the program does not take into the account any elastic properties of the pile.

The acceleration at the pile toe is calculated by:

$$a_{toe} = (F_c - SRV_f)/m_{td} \qquad (6)$$

Where $m_{td}$ is the total dynamic mass (dynamic weight of vibratory hammer, clamp(s), and pile).

The displacement amplitude (top-top) is calculated similarly using:

$$d_{toe\,(t-t)} = 2.(F_c - SRV_f)/(m_{td} . \omega^2) \qquad (7)$$

For a situation without soil resistance formula (7) is equal to the well-known formula $a = 2.M_e/m_{td})$.

The calculations are stopped when one of the following criteria are met:

- final penetration has been reached
- $SRV_{toe}$ is larger than $F_{toe}$
- $d_{toe\,(t-t)}$ is smaller than a given minimum amplitude

The minimum given amplitude should be based on the expected elastic behavior of the soil under vibratory loading conditions.

The $\beta_{min}$-values are used as input in the program. The program calculates the actual β-values based on the toe acceleration in the previous step.

The $\beta_{min}$-values as well as the minimum amplitude are given per soil layer, i.e. at the same total resistance to driving a pile might meet refusal in a sticky clay layer but continue penetration into a sand layer.

The program has an option to include a thin soil layer around the pile in the mass and acceleration calculations. This might be necessary e.g. to better simulate a pile which is extracted from a clay type soil. In many cases the failure mode in these situations is clay-to-clay instead of clay-to-pile.

The program does not have a feature to vary the frequency during penetration. Normally the frequency drops when operating pressures increase, i.e. generally drops with increasing penetration. To incorporate such a feature in a proper way would require complicated power balance calculations. For normal predictions generally a lower frequency was assumed than the maximum operating frequency of the hammers.

## 5 PREDICTION TRANSVIB-CONFERENCE.

### 5.1 General information-Background

To investigate the strength, accuracy and sensitivity of several vibratory prediction methods, a contest was organized by the organizing committee of the First International Conference on Vibratory Pile Driving and Deep Compaction. The test itself was carried out at a site in Montoir, Nantes, France. Details of the test are presented in a different paper of this conference and are not repeated here.

Only for completeness and legibility of this paper when read independently from the procedures, are the most important soil information given below.

The soils consists of a loose sandy top layer of 2 meter thickness (cone-resistance value $q_c$ average of 3 Mpa) followed by a very dense sand deposit down to 8 m penetration with average $q_c$ of 12-18 Mpa. Below this sand layer the soil consist mainly of alternating layers of clay and sand, whereby the sand layers are predominant down to approximately 16m, a 50/50 percent presence to 23m and below this level mainly clay is encountered.

The driven test piles had a diameter of 339mm and a wall thickness of 14mm. The pile had a length of 32m. A U-profile was welded to both sides of the piles to protect the accelerometers and strain gauges, which were connected at several levels to the pile.

An ICE-815 (low frequency) vibratory hammer was used to drive the piles. A pipe-pile clamp was bolted to hammer to grip the piles. In this configuration the ICE-815 had an eccentric moment of 46 kgm, a vibrating weight of 5500 kg and a total weight of 8000 kg. The maximum frequency of the hammer is 1570 cpm (26,2 Hz). During the tests the actual frequency varied from 1320 cpm to 1560 cpm (20 to 26 Hz). The higher frequency was probably generated during the early stages of driving and decreased slowly with increasing operating pressures at higher penetrations.

Corresponding centrifugal forces $F_c$ at 1320 and 1560 cpm are respectively 884 kN and 1235 kN.

For the prediction event in spring 2001 the dimensions of the pile itself only were given, and predictions had to be made for operating frequencies of 800 cpm (13 Hz), 1200 cpm (20 Hz) and 1600 cpm (26,7 Hz). During actual installation hammer was operated only at full speed. For this reason are only the graphical results for this case are presented in this paper. (The outcome of the other two situations were predicted as refusal on top of the dense sandlayer for a frequency of 800 cpm and a critical equilibrium for 1200 cpm, i.e. the pile could meet refusal but also punch through the sand layer).

### 5.2 Prediction results-input data as given prior to tests

Based on previous experience it was expected that the maximum operating frequency of the hammer would not exceed 1450 cpm (24,2 Hz) at high operating pressure. For the prediction that frequency was therefore selected. β-values were selected on

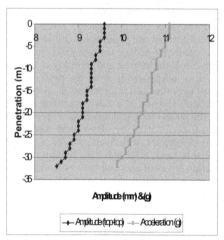

Figure 3. Decrease in acceleration and displacement amplitudes

the experience with mainly low-displacement piles only. It was further assumed that the footplate underneath the pile was larger in diameter than the pile itself. This was taken from presented data of a test with a D-19 diesel hammer on the same site in 2000. A large footplate will act as 'shoe' or as 'friction reducer' and result in lower β-values. The results of the first prediction calculations are given in figures 4 at the end of this paper.

Figure 3 shows the calculated decrease in pile amplitude (top-top) and acceleration. Noticeable is the stronger decrease in the clay layer below 22 meters by reason of higher β-values.

From figure 4 it can be seen that the available driving force at the toe of the pile $F_{toe}$ exceeds at any depth the resistance $SRV_{toe}$. In the prediction stage of the contest it was therefore expected that the ICE-815 would be capable to drive the pile to full penetration. It was however mentioned that thin sandlayers exceeding 3 times the diameter, i.e. exceeded 1-meter thickness could cause early refusal.

### 5.3 Prediction results – factual input data

When the results of the tests were made available, a new prediction was made using the same soil parameters.

The changes between the first prediction run and the second covered:
- a larger pile toe and friction area because of the two U-profiles welded at the side of the pile for protection of the strain gauges and accelerometers.
- - the assumption of a thicker sand layer at 18 m. The cone resistance for this layer was taken as 10 MPa.

The results of the calculations are shown in Fig 5.

Without any modification to the β-values, the program predicts that full penetration will still be reached, although the available driving force and resistance at the toe are much closer together than in the first predictions.

In case the minimum operating frequency of 1320 cpm (corresponding $F_c$ equal 884 kN) the right hand line on Figure 5 shifts nearly 200 kN to the left. As the program does have to feature to vary the frequency during one run, the most likely decrease for the drive force at the toe is from 1066 kN at the service (operating frequency 1450 cpm) to 800 kN at 20m penetration. (Operating frequency 1320 cpm).

### 5.4 Postdiction results – factual input data

To match predictions and factual driving conditions the following input changes in the program were made:
- increase the β-values for friction to approximate double their original value. This was considered

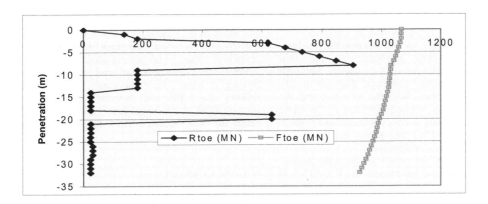

Figure 4. Prediction Results based on original given test data.

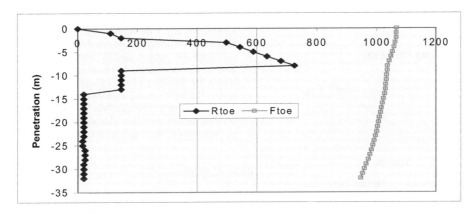

Figure 5. Prediction Results based onfactual data.

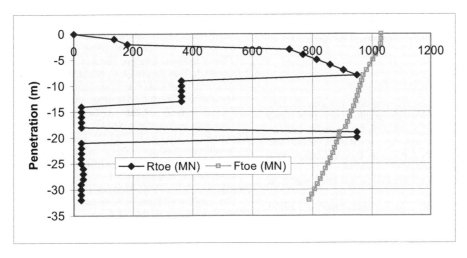

Figure 6. Postdiction results based on modified input data.

acceptable because they were original based on the experience with sheet piles and open ended pile piles. It was further assumed in the first predictions that the piles had an oversized foot plate decreasing the friction and thus the β-values.

− Increase the cone resistance in the thin intermediate layer from 18-20 meter. Although the cone penetration test itself does not rectify such an assumption, both the SPT test and the PMT test results showed densities in the same order of magnitude as the dense sandlayer from 2-8 meter. The PMT test shows even a higher Young's modulus for this layer.

Figure 6 shows the results of the calculations. With this set of input parameters the ICE-815 will drive the pile through the upper layer, however just, but will meet refusal at 18 m. At this level is the driving force $F_{toe}$ smaller than the resistance $SRV_{toe}$.

Figure 7 shows the decrease of the available driving force at the toe for all three cases in one graph.

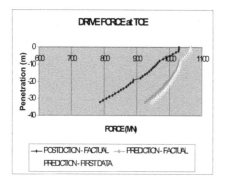

Figure 7. Drive force at toe for all 3 cases.

To take into account the effects of the decreasing frequency with depth the lines would start at the same point at the surface but would be 200 kN lower at 20 m penetration.

### 5.5 Rate of penetration - Timecount

The ICE β-method program does not give the rate of penetration. Although graphs (not published) exists on the rate of penetration based on the ratio of centrifugal force and toe resistance, these are considered not very reliable as they are based on a few back analyses only.

For this reason no prediction has been made on the rate penetration.

Jonker (Jonker and Middendorp, 1988) introduced the term Timecount as a measure for the

rate of penetration. The timecount (in seconds/meter) is the reciproke value of the common used rate of penetration expressed in meters/minute. The advantage of the timecount is that it is similar to the blowcount registration with impact hammers. It also gives high values (it takes more seconds to penetrate 1 meter) when driving is hard, and low values when driving is easy. The 'timecount' lines generally follows the CPT and N-value lines out soil investigations. A timecount line is therefore visually easier for interpretations than a rate of penetration line.

## 6 CONCLUSIONS

The following conclusions can be made from the predictions and postdiction analyses:

1. The ICE β-method is still a valuable, powerful and easy tool to predict the vibrateability of piles.
2. Calibration and validation of the β-factors is very important for accurate and reliable predictions.
3. β-factors are approximately double for closed ended piles compared to open ended piles. The larger the diameter of the closed-ended pile, the higher the increase in β-factors is expected to be.
4. Intermixed soils whereby thin layers of sand are interbedded in a mainly clay deposit complicate the accuracy of the calibration of the program.

## REFERENCES

Jonker G. 1987. Vibratory Pile Driving Hammers for Pile Installations and Soil Improvements Projects. *OTC 5422 Proceedings Offshore Technology Conference, Houston USA.*.

Jonker G., Middendorp P., 1988. Prediction of Vibratory Hammer Performance by Stress Wave Analyses. *3rd International Stress Wave Conference on the application of Stress Wave Theory to Piles. Ottawa, Canada*

Ligterink A, Van Zandwijk C, Middendorp P., 1990. Accurate Vertical Pile Installation by Using a Hydraulic Vibratory Hammer on the Arbroath Project. *OTC 6236 Offshore Technology Conference, Houston, USA.*

Ligterink A.,Martin R., 1992. Field Experience Installing A-shape Piles with a Vibratory hammer. *OTC 6842. Offshore Technology Conference. Houston, USA.*

Warrington D.C., 1990. Methods For Analysis of the Driveability of Piles by Vibration. *Article in Pile Buck January and March Issue.*

*Vibratory Pile Driving and Deep Soil Compaction - TRANSVIB2002,*
*Holeyman, VandenBerghe & Charue (eds.), © 2002 Swets & Zeitlinger, Lisse, ISBN 90 5809 521 5*

# Class A Prediction of the Vibro-Drivability of a Steel Pile in Montoir

Roberto O.Cudmani
*Institute of Soil and Rock Mechanics, University of Karlsruhe*

A class "A" prediction of the vibro-drivability of a steel pile is presented. The numerical model, the calibration of the parameters, the assumptions made for the numerical calculations are summarized and the most relevant results are briefly presented and discussed. Differences between the actual pile geometry and machine parameters with respect to the values used in the numerical modelling, but also a proposed simplification of the cone resistance profile cause the discrepancy between calculated and experimental results.

## 1 INTRODUCTION

In the framework of the French national project on vibratory driving, the behavior of vibratory driven piles in the harbor of Montoir, located at the estuary of the Loire River was investigated in large scale driving tests. Two 339 mm closed-end tubes, 31 m long, were installed by vibratory driving. The subsoil consists of alluvial sandy deposits overlying mud clay. The test site, the site investigation program, the instrumentation and installation of the piles as well as the most relevant results of the investigation are summarized in BOREL et al. (1). The numerical model, the calibration of the parameters, the assumptions made for performing the class "A" prediction as well as the most relevant results are presented and discussed in the next sections.

## 2 NUMERICAL MODEL

The vibro-drivability prediction was carried out using the mechanical model for the description of vibro-driving of piles published in a companion paper (CUDMANI et al. (3)). The model is based on the wave propagation method. The pile is assumed to behave as an elastic rod. The soil resistance is divided into shaft and tip resistance and its evolution is modelled by mathematical functions, which are based on experimental results and justified from a soil mechanics point of view. The solution of the non-linear second order differential equation describing the problem is obtained numerically using a finite element scheme. The finite element model and the adopted boundary conditions are schematically shown in Figure 1.

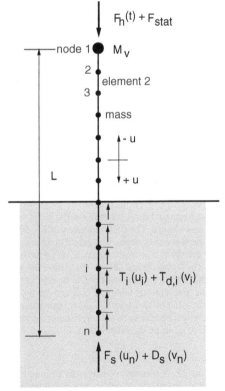

Figure 1. FE-Model used for the numerical simulation of pile driving.

## 3 MODELLING OF THE PILE

The pile was assumed to be an elastic closed-end rod. 129 one dimensional two-node elements were used for the discretization of the pile. The geometry and material properties adopted in the calculations are:

| | |
|---|---|
| pile length | 32 m |
| pile diameter | 339 mm |
| tube thickness | 14 mm |
| density | 7.860 t/m$^3$ |
| toe area | 0.09026 m$^2$ |
| Elastic Modulus | $2.1 \cdot 108$ kN/m$^2$ |

## 4 MODELLING OF THE DRIVING FORCE

The driving force is assumed to be harmonic:

$$F_h = F_{stat} + S_v(2\pi f)^2 \sin(2\pi f t) \qquad (1)$$

The calculations were carried out for f=13.3, 20 and 26.6 Hz (800, 1200 and 1600 rpm). The static component of the driven force was assumed to amount the total weight of the vibrator (85.5 kN). The eccentric moment was $S_v$=46 kgm. The vibrating mass $M_v$=5050 kg was attached to the top of the pile.

## 5 EVALUATION OF THE SOIL RESISTANCE

The evaluation of the parameters for modelling of toe and shaft resistance was done on the basis of the cone penetration resistance qc. A mean cone resistance per meter qm was determined in those depths in which the numerical calculations were performed.

### 5.1 Parameters of the Shaft Resistance

In order to determine the shaft resistance, the soil profile was simplified in two layers. The top layer includes the fill, the coarse sand, gravely sand and gravely clayey sand layers. The bottom layer consists of the mud clay. In the top layer the maximum shear stress was calculated with the expression:

$$\tau_{max} = K\sigma_v \tan\delta \qquad (2)$$

For the horizontal earth pressure coefficient a value of $K = 0.5$ was adopted. The friction angle was $\delta = 30°$. The vertical pressure over depth was calculated as $\sigma_v = \gamma h$. The dry density was estimated from $q_m$ using the method proposed by CUDMANI and OSINOV (2). Above the ground water table, the soil was considered to be dry, i.e $\gamma = \gamma_d$. Below the ground water table, the buoyant density $\gamma = \gamma'$ was used. In the bottom layer the maximum shear stress was $\tau_{max} = c_u$. The undrained shear resistance cu was estimated from $q_m$ using the relationship:

$$c_u = \frac{q_m - \sigma_v}{N_k}, \qquad (3)$$

with $N_k$=17. A constant density was assumed in the bottom layer in order to calculate the vertical pressure to be used in refcu. The estimated distribution of maximum shear stress over depth is shown in Figure 2.

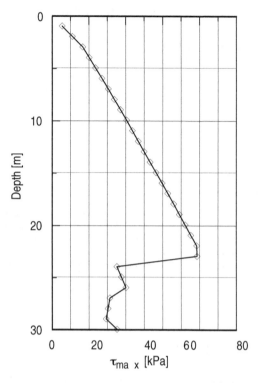

Figure 2. Estimated distribution of maximum shear stress over depth.

The constant C1 defining the shape of the shear stress/strain-relationship was calculated with the expression (CUDMANI et al. (3)):

$$C_1 = \frac{\ln(1 - 0.95)}{0.01\ 0.339} \tau_{max} \qquad (4)$$

### 5.2 Modelling of the Toe Resistance

For modelling the toe resistance, four parameters, namely $C_b$, $\beta$, $C_e$ and $q_s$ were evaluated. In order to determine Cb, the cone penetration resistance was converted to the penetration resistance $N_{10}$ of a Dynamic Probing (DPH) by means of the empirical relationship $N_{10} = q_m$ [MPa]. Then, $C_b$ was calculated using the empirical function $C_b$ [kPa] = 1000 $(N_{10})^2$. The parameter $\beta$ was estimated on the basis of the experimental values for sand and silty sand found in large scale vibro-driving tests by CUDMANI et al. (3). $\beta$ varied between 0.4 (for $q_m > 12$) and 0.6 (for $q_m < 4$). The parameter $C_e$ does not have a great influence on the numerical results, for the calculations 4 $C_b$ < Ce < 5 $C_b$ was assumed. The static penetration resistance was assumed to be equal to the cone penetration resistance $q_s = q_m$. Figure 3 shows the variation of the model parameters $C_b$ and $\beta$ over depth.

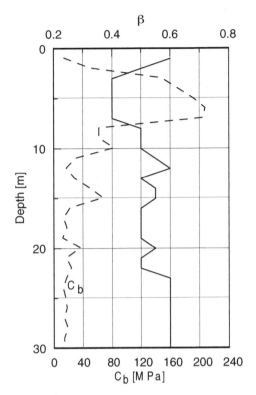

Figure 3. Distribution of the parameters $C_b$ and $\beta$ over depth.

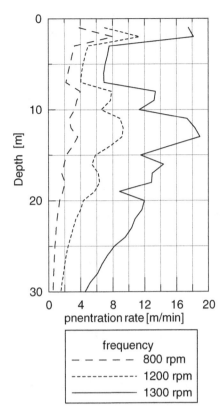

Figure 4. Predicted penetration rates over penetration depth.

## 5.3 MODELLING DAMPING DUE TO WAVE RADIATION

In order to model damping due to wave radiation the shear stiffness for small deformations $G$ was estimated using the empirical relationship between $G$, void ratio $e$ and mean pressure $p$ provided by Seed and Idriss (5) for cohesionless and by HARDIN and DRNEVICH (4) for cohesive soils.

## 6 RESULTS AND DISCUSSION

In this section prediction results are presented and briefly discussed. Reference is done to the results of the full-scale test given by BOREL et al. (1).

Figure 4 shows the predicted penetration rates for three different vibrator frequencies. Qualitatively, calculated and measured penetration rates show approximately the same change with penetration depth. However, the numerical model overestimated the recorded penetration rates and was not able to predict pile refusal.

There are five main reasons for the overestimation of the penetration rates:

1. The discrepancy between the actual and assumed toe surface. The actual toe would lead to an increase of the penetration resistance and to a reduction of the penetration rate.
2. The discrepancy between the actual and the assumed vibrator frequency. The calculations were carried out for 800, 1200 and 1600 rpm. The actual vibrator frequency varied between 1300 and 1400 rpm.
3. A overestimation of the static component of the driving force, which in the model was assumed to be the total weight of the vibrator (vibrating mass and bias weight).
4. The CPT results does not show any increase of the cone resistance at/above the refusal depth as PMT and SPT do.
5. An underestimation of the penetration resistance due to the use of the mean cone resistance $q_m$ instead of $q_c$ for the determination of the model parameters.of 10 m depth.

Figure 5 compares the original, mean and maximum cone resistance with the PMT-limit pressure and the SPT-blow count. The maximum cone resistance $q_{c,max}$ is the maximum value of $q_c$ per meter. As it can be seen

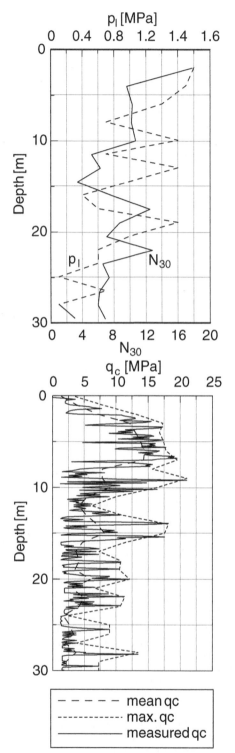

Figure 5. Comparison of CPT, PMT and STP results.

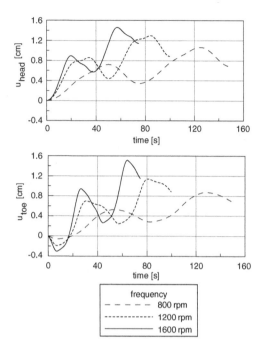

Figure 6. Predicted displacement time histories at the pile head (top) and toe (bottom) at a penetration

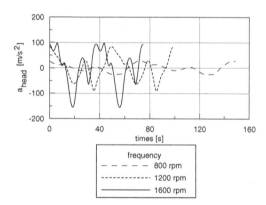

Figure 7. Predicted acceleration time histories at the pile head at a penetration of 10 m depth.

due to the strong variations in the original CPT-data, qm deviates very strongly from the maximum cone resistance $q_{c,max}$ which correlates better with the results of PMT and SPT.

A better agreement of actual and predicted penetration rates could be obtained if pl or N30 or the maximum cone resistance would have been used for the determination of the model parameter.

Figure 6 shows the time history u(t) of the predicted displacement at the pile head and the pile toe. In this

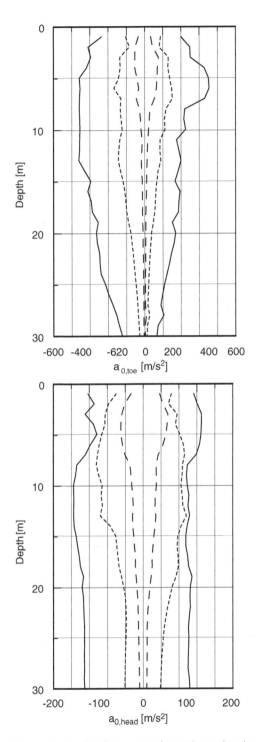

case the prediction was more satisfactory, though the discrepancies between model and reality mentioned above also affect the displacement time histories.

Figure 7 compares the measured and predicted acceleration time histories at the head and at the toe of the pile. Figure 8 shows the distribution of the acceleration amplitudes at the pile head and pile toe over penetration depth. The acceleration amplitude was estimated as:

$$a_0 = \frac{a_{max} - a_{min}}{2} \qquad (5)$$

Here, $a_{max}$ and $a_{min}$ are the maximum downward and upward acceleration amplitudes in the considered penetration depth. Prediction match measurement quite well since the mentioned discrepancy sources with exception of the vibrator frequency do not affect acceleration time histories very strongly.

REFERENCES

1. S. Borel, L. Gianeselli, D. Durot, P. Vaillant, L. Barbot, B. Marsset, P. Lijour. The Full-Scale Behaviour of Vibratory Driven Piles in Montoir. *Proceedings*: International Conference on Vibratory Driving and Deep Compaction, 2002 (in this volume).

2 R. Cudmani, V. Osinov: The Cavity Expansion Problem for the Interpretation of Cone Penetration and Pressuremeter Tests. *Canadian Geotechnical Journal*, **38**, pp. 622–638, 2001.

3 R. Cudmani, G. Huber, G. Gudehus: A Mechanical Model for the Investigation of the Vibro-Drivability of Piles in Cohesionless Soils. *Proceedings*: International Conference on Vibratory Driving and Deep Compaction, 2002 (in this volume).

4. B.O. Hardin, V.P. Drnevich. Shear Modulus and Damping in Soils: Design Equations and Curves. *Journal of Soil Mechanics and Foundations Divisions*, ASCE, Vol. 98, **SM7**, pp.667–692, 1972.

5. H.B. Seed, I.M. Idriss. Soil Moduli and Damping Factors for Dynamic Response Analyses. *Report EERC 70–10*, Earthquake Engineering Research Center, University of California, Berkeley, 1970.

Figure 8. Predicted vs. experimental acceleration amplitudes at the pile toe and pile head over penetration depth.

*Vibratory Pile Driving and Deep Soil Compaction - TRANSVIB2002,*
*Holeyman, VandenBerghe & Charue (eds.), © 2002 Swets & Zeitlinger, Lisse, ISBN 90 5809 521 5*

# Vibrodrivability of piles at Montoir prediction using a simplified model

J.-G. Sieffert
*ENSAIS – IMFS UMR 7507 CNRS-ULP, Strasbourg, France*

ABSTRACT: For the prediction of vibrodrivability of piles at Montoir the simplified model presented at the International Conference TRANSVIB 2002 was used. These predictions indicate the vibrodriving until a depth of 30 meters is only possible with the highest frequency (1600 rpm) of the vibrator. The pile penetration will be difficult through the coarse sand layer (2 to 7 meters depth).

## 1 ASSUMPTIONS

For the prediction of vibrodrivability of piles at Montoir a simplified model was used. The most important assumptions are:
- the pile is considered as a rigid body,
- the behaviour law for the toe resistance and the shaft resistance is rigid plastic, without "quakes" or viscosity effects.

The complete analysis of this model is developed by Sieffert (2002).

## 2 VIBRATOR AND PILE FEATURES

For practical reasons (reduction of bending deformation of the pile), it was planned the pile will be vibrodriven in three steps. In the first step, the length of the pile is 12. 51 m and will be driven until 11 meters depth. In the second step, an additional tube (length: 12.16 m) is fixed at the first pile, so that the new length of the pile is 24.67 m and will be driven until 23 meters depth. In the last step, a new tube (length: 12.33 m) is added and the last length of the pile length will be 37 m. Therefore we have to consider these three lengths $L$ of the pile. Table 1 gives the corresponding vibrating masses $M_v$ and static masses $M_s$ of the system vibrator-clamp-pile.

Table 1. Vibrating and static masses

| $L$ (m) | $M_v$ (kg) | $M_s$ (kg) |
| --- | --- | --- |
| 12.51 | 7 477 | 9 927 |
| 24.67 | 8 864 | 11 314 |
| 37.00 | 10 270 | 12 720 |

For these three lengths, the displacement amplitude (0 to peak) at the top and at the bottom were calculated considering the system totally free in two cases: the pile is an elastic body and the pile is a rigid body. The results are given in table 2.

Table 2. Displacement amplitudes at the top – at the bottom, and for a rigid body (mm)

| $L$ (m) | at the top | at the bottom | rigid body |
| --- | --- | --- | --- |
| 12.51 | 6.08 | 6.62 | 6.15 |
| 24.67 | 4.76 | 6.83 | 5.19 |
| 37.00 | 3.05 | 8.41 | 4.48 |

We see that the displacement amplitude at the top decreases and the displacement amplitude at the bottom increases with the length of the pile for the elastic body. In the same time, the amplitude is decreasing for the rigid body.

We have also calculated the force ratio (amplitude of the dynamic force at the top of the pile divided by the amplitude $F_{exc}$ of the excitation force). The numerical results are given in table 3.

Table 3. Force ratio (%)

| $L$ (m) | elastic body | rigid body |
| --- | --- | --- |
| 12.51 | 20.0 | 19.1 |
| 24.67 | 37.4 | 31.8 |
| 37.00 | 59.9 | 41.1 |

It is clear that this force ratio increases with the pile length. On the other hand, we see that the results concerning an elastic body and a rigid body (Tables 2 and 3) are close for the two first lengths of the pile, but not for the third length. We can conclude that the assumption of a rigid body is acceptable for the two first lengths, and probably not very pertinent for the third length.

## 3 SHAFT AND TOE RESISTANCES

The shaft and the toe resistances are deduced from the soil analysis (Simecsol, 1999).

### 3.1 Shaft resistance

The soil analysis report gives the effective volume weight and the internal friction angle as a function of the depth (Table 4).

Table 4. Effective volume weight, internal friction angle

| depth (m) | $\gamma'$ (kN/m3) | $\phi$ (degrees) |
|---|---|---|
| 0 - 3 | 19 | 20 |
| 3 - 10 | 11 | 35 |
| 10 - 22 | 11 | 25 |

With these values, it is easy to calculate the static normal earth pressure applied on the pile applying the classical methods using the earth pressure coefficients:

$$K_{a\gamma} = K_{aq} = \tan^2\left(\frac{\pi}{4} - \frac{\phi}{2}\right) \quad (1)$$

This earth pressure is given on figure 1 versus depth.

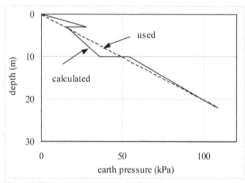

Figure 1. Earth pressure versus depth

In order to simplify the calculation, we will use a linear function for this earth pressure:

$$\sigma = 4.91\,z \quad (2)$$

where the stress $\sigma$ is given in kPa and the depth $z$ in meters. We obtain the normal resulting static earth force $F$ using:

$$F = 0.5\,s\,\sigma\,z \quad (3)$$

in which $s$ is the perimeter of the pile (1.07 m).

At the end we suppose that the resulting dynamic shaft resistance $F_l$ is a part (15%) of the normal static earth pressure.

$$F_l = 0.4\,z^2 \quad (4)$$

where $F_l$ is given in kN and $z$ in meters.

### 3.2 Toe resistance

For the dynamic toe resistance, we use the static cone resistance $qc$ obtained by the Cone Penetration Test (CPT). Figure 2 shows the measured values of the cone resistance versus depth.

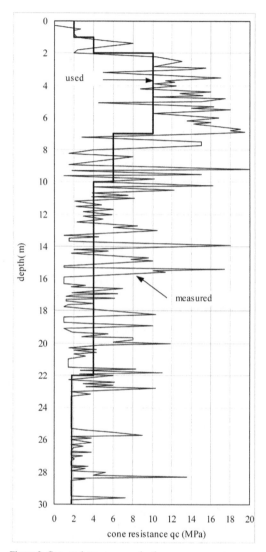

Figure 2. Cone resistance versus depth

Table 5. Cone and toe resistances versus depth

| depth (m) | qc (MPa) | $F_l$ (kN) |
|---|---|---|
| 0 - 1 | 2 | 181 |
| 1 - 2 | 4 | 362 |
| 2 - 7 | 10 | 1 087 |
| 7 - 10 | 6 | 544 |
| 10 - 22 | 4 | 362 |
| 22 - 31 | 1.8 | 161 |

In order to simplify the calculations, we will consider constant values for the cone resistance $q_c$ and the toe resistance $F_p$ (see table 5 and figure 2).

## 4 RESULTS FROM THE ANALYSIS

The calculations were performed with aid of the software BRAXUUS. No external static retaining force was considered. Only the most important results are presented here.

### 4.1 Penetration curve and penetration speed

The penetration curve (mean penetration depth versus time) is given on figure 3.

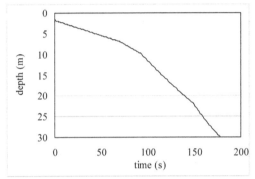

Figure 3. Penetration curve

The penetration speed (mean penetration depth versus mean penetration speed) is given on figure 4.

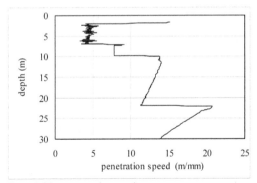

Figure 4. Mean penetration speed

These figures show clearly the effect of the toe resistance on the mean penetration speed:
- each change of the toe resistance gives an important change of the mean penetration velocity,
- the smallest penetration speed is obtained for the largest toe resistance (1 087 kN from 2 till 10 m): the model results indicate that the pile penetration

will be difficult through this layer and that the vibrating behaviour will be more or less chaotic.

We can see also the decreasing of the penetration speed with the depth when the toe resistance is constant (see for example from 10 till 22 m, and from 22 till 30 m). That describes directly the effect of the increasing of the shaft resistance with the penetration depth for a constant toe resistance in the same time.

On the other hand, the increase of the force ratio (19.1 to 31.8% and 31.8 to 41.1%) due to the length change of the pile (respectively at 11 and 23 m depth) has not a significantly effect on the mean penetration speed.

### 4.2 Peaks downward and upward of the acceleration

Figure 5 gives the peak downward, peak upward and the RMS values of the acceleration versus depth. The important change of these curves at 11 and 23 m depth can perhaps be explained by the change of the pile length and consequently the change of the value of the vibrating mass. On the other hand, these accelerations decrease even when the mean penetration speed increases in the same time.

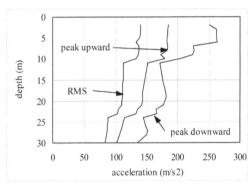

Figure 5. Acceleration versus depth

### 4.3 Alternative movements of the pile

We present here the results concerning the alternative movements of the pile at four depths: 5, 10, 20 and 30 meters. Figures 6, 7 and 8 show respectively the displacement, the acceleration and the velocity during four periods of the excitation force. It is clear that these three functions are periodic. The period is exactly the period of the excitation force at the depths 10, 20 and 30 meters, and the double of the excitation force period at the depth 5 meters. But in all cases the relative peaks are separated by a period of the excitation force.

Table 6 shows clearly that the resistance force is the toe resistance (the shaft resistance can be neglected) at depth 5 meters.

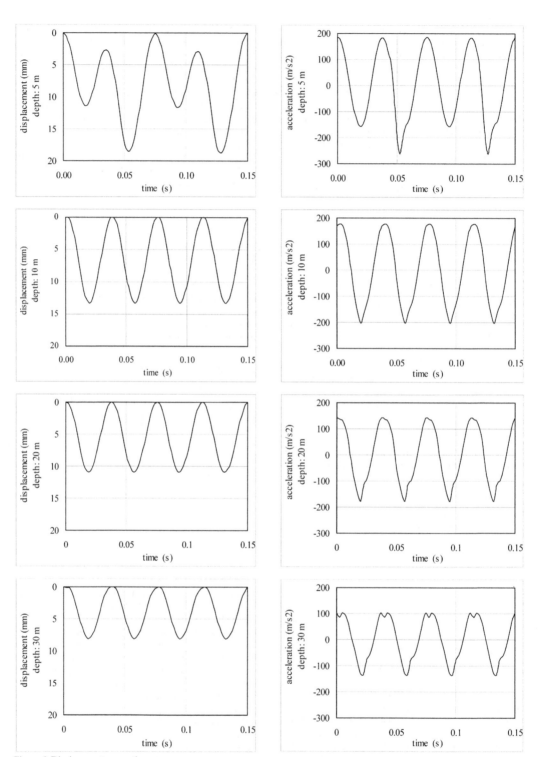

Figure 6. Displacement versus time

Figure 7. Acceleration versus time

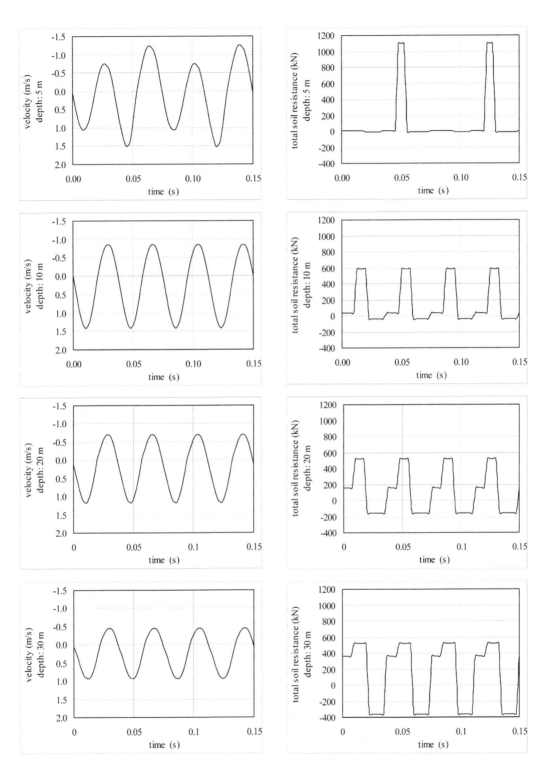

Figure 8. Velocity versus time

Figure 9. Total soil resistance versus time

Table 6. Shaft and the toe resistances

| depth (m) | $F_l$ (kN) | $F_p$ (kN) |
|-----------|-----------|-----------|
| 5 | 10 | 1 087 |
| 10 | 40 | 544 |
| 20 | 160 | 362 |
| 30 | 360 | 161 |

In order to better understand this behaviour we analyse yet the total soil resistance.

### 4.4 Total soil resistance

Figure 9 shows that the period of the total soil resistance is the double of the excitation force period at 5 meters depth, and the excitation force period at 10, 20 and 30 meters depth. Figures 10 give the depth versus time and confirm this result. In the first case, the penetration depth is only larger during the second period after the largest previous penetration depth. The contact of the toe on the soil is very short and the toe resistance occurs only every other excitation force period. At a 10 meters penetration depth, the contact of the toe on the soil is larger and the toe resistance occurs every excitation force period.

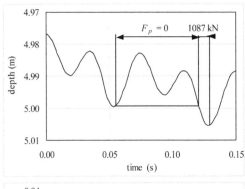

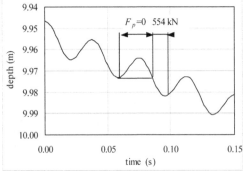

Figure 10. Penetration depth versus time

It is also interesting to compare the amplitudes of the alternative movements and the mean penetration velocity (see table 7).

Table 7. Mean penetration speed – alternative movements amplitude

| depth (m) | penetration speed (m/s) | displacement (mm) | velocity (m/s) | acceleration (m/s2) |
|-----------|-------------------------|-------------------|----------------|---------------------|
| 5 | 0.07 | 6.7 | 1.3 | 192 |
| 10 | 0.13 | 6.6 | 1.1 | 187 |
| 20 | 0.20 | 5.4 | 0.9 | 158 |
| 30 | 0.23 | 4.1 | 0.7 | 118 |

The amplitude of the alternative velocity is larger than the mean penetration speed for the four considered penetration depths. The increasing of the mean penetration speed and the decreasing of the alternative movement amplitudes with the penetration depth can be explained here by the increasing of the shaft resistance (alternative movement amplitudes) and the decreasing of the toe resistance (mean penetration speed) with the penetration depth. But the Figure 4 shows that the shaft resistance has also an effect on the mean penetration speed.

### 4.5 Shaft resistance – toe resistance diagramme

We introduce dimensionless parameters as following:

$$X = \frac{F_l}{P_s} \qquad Y = \frac{F_p}{P_s} \qquad A = \frac{F_{exc}}{P_s} \qquad (5)$$

$P_s$ is the static weight of the system vibrator-clamp-pile. The values corresponding at the four considered depths and the "static" lines are given on figure 11. The static lines are obtained by writing that the sum of the amplitudes of the resistance forces equals the sum of the amplitudes of the active forces (equation 6).

$$X + Y = A + 1 \qquad (6)$$

This writing is valid for static forces but, of course, this equation does not make sense for dynamic motions and is given here only as information. Nevertheless the point "5m" corresponds to the highest value of the toe resistance and is in the same time the nearest to the corresponding "static" line.

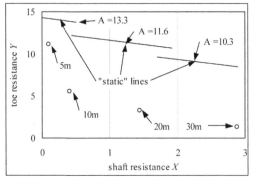

Figure 11. Diagramme $X$ - $Y$

230

# 5 CONCLUSIONS

Our predictions of vibrodrivability of piles at Montoir based on a simplified modelling for the pile and for the both shaft and toe resistances indicate that the pile penetration will be difficult through the coarse sand layer (2 to 7 meters depth) and only possible using the highest excitation force available from vibrator (1600 pm). The shaft resistance can be niggled regarding the toe resistance in this layer. The alternative movements will be more impacts movements than vibrations movements. Our model (rigid) is not able to describe the compression waves induced by these impacts in terms of displacements, velocities and accelerations but can give interesting indications for a better knowledge of the dynamic soil-pile interaction during vibrodriving.

## ACKNOWLEDGEMENT

The model used for this penetrability prediction was carried out in the framework of the French national project on vibro-driving. This project is managed by IREX and forms part of the operations of the RCG&U network. The French Ministry of Public Works, which is acknowledged for its financial support, sponsored it.

## REFERENCES

Sieffert, J.-G. 2002. Vibratory pile driving analysis. A simplified model. *Int. Conf. on Vibratory Pile Driving and Deep Soil Compaction, TRANSVIB 2002, Louvain-la-Neuve, 9-10 Sept.*
Simecsol. 1999. Port de Montoir T.M.D.C. 4 – Campagne géotechnique et instrumentation en cours de battage d'un pieu d'essai, *Note de présentation.*

*Vibratory Pile Driving and Deep Soil Compaction - TRANSVIB2002,*
*Holeyman, VandenBerghe & Charue (eds.), © 2002 Swets & Zeitlinger, Lisse, ISBN 90 5809 521 5*

# Author index

Baek, Y.I. 175
Barbot, L. 179
Baycan, S. 83
Berghe van den, J.F. 61, 193
Borel, S. 167, 179
Bustamante, M. 167

Cudmani, R.O. 45, 157, 219

Durot, D. 179

El-Mossallamy, Y. 123

Fellin, W. 115
Geiß, A. 115
Gianeselli, L. 167, 179
Gudehus, G. 45

Haegeman, W. 135
Hochenwarter, G. 115
Holeyman, A. 3, 61, 89
Huber, G. 43, 157
Huybrechts, N. 89

Jonker, G. 213

Kissel, W. 123
Kwon, S.Y. 175

Lee, J.H. 175
Legrand, C. 89
Lijour, P. 179
Löschner, J. 123

Madabhushi, S.P.G. 147
Marsset, B. 179
Meijers, P. 141

Rainer Massarsch, K. 33
Raju, V.R. 129
Rausche, F. 21

Schlesinger, B. 123
Seo, D.D. 175
Sieffert, J.G. 53, 225

Thusyanthan, I. 147
Tol van, A.F. 141

Vaillant, P. 179
Vié, D. 69
Viking, K. 99

Wehr, W.C.S. 129
Won, J.H. 175

Zouw van der, N. 213

T - #0266 - 101024 - C0 - 254/178/13 [15] - CB - 9789058095213 - Gloss Lamination